TERRITORIES OF URBANISM

PAOLA VIGANÒ

TERRITORIES OF URBANISM

THE PROJECT
AS KNOWLEDGE PRODUCER

TRANSLATED FROM THE ITALIAN BY STEPHEN PICCOLO

EPFL Press

A Swiss academic publisher
distributed by Routledge

Routledge
Taylor & Francis Group
www.routledge.com/builtenvironment

Taylor & Francis Group Ltd
2 Park Square, Milton Park Abingdon,
Oxford, OX14 4RN, UK

Routledge is an imprint of the Taylor & Francis Group,
An informa business.

Simultaneously published in the USA and Canada by Routledge,
711 Third Avenue, New York, 10017

www.routledge.com

Library of Congress Cataloging-in-Publication Data
A catalog record for this book is available from the Library of Congress

ÉCOLE POLYTECHNIQUE
FÉDÉRALE DE LAUSANNE

The author and publisher express their thanks to the
Ecole polytechnique fédérale de Lausanne
for its generous support towards the publication of this book.

This book is published under the editorial direction of professors Luca Ortelli and Jacques Lucan.

Coordination of the English version: Martina Barcelloni Corte

Proofreader: Emily Lundin

EPFL Press

The EPFL Press is the English-language imprint of the Foundation of the Presses polytechniques et universitaires romandes (PPUR). The PPUR publishes mainly works of teaching and research of the Ecole
polytechnique fédérale de Lausanne (EPFL), of universities and other institutions of higher education.

Presses polytechniques et universitaires romandes
EPFL – Rolex Learning Center
Post office box 119
CH-1015 Lausanne, Switzerland
E-mail: ppur@epfl.ch

www.epflpress.org

© 2016, First edition, EPFL Press
ISBN 978-2-940222-89-6
Printed in Italy

Original title: *I territori dell'urbanistica. Il progetto come produttore di conoscenza*,
© Officina Edizioni, Rome, 2010

Table of Contents

Dedicated to Alberto Viganò

There are many people I would like to thank.

Cristina Bianchetti for her reading, apt recommendations and constant support;

Phyllis Lambert, Mirko Zardini and Alexis Sornin for the intense period of study at CCA and the great courtesy of its archivists and librarians;

Stefano Peluso for the translations from German and for his comments;

Lorenzo Fabian, Laura Domenichini and Giovanni De Roia who made the final (and therefore fundamental) contribution to the completion of the book, as well as Simona Bodria, Alessia Calò, Dao-Ming Chang, Uberto degli Uberti, Tommaso Fait, Steven Geeraert, Alvise Pagnacco, Kasumi Yoshida of the studio Secchi-Viganò, who took part in its slow construction.

Thanks also to Tullia Lombardo and Giulia Fini, assistants in the course on Techniques of Urbanism and Giambattista Zaccariotto, Irene Guida, Stefano Peluso, Monica Bianchettin, who collaborated on the Urban Design Workshop at IUAV, University of Venice, for their precious help.

I would also like to thank the students of the schools of architecture of Lausanne (EPFL), Louvain (KU Leuven), Aarhus, and of the European Master in Urbanism (EMU), and above all my students at IUAV in Venice, who in recent years have participated with imagination and enthusiasm in the definition of the main hypotheses of the book, and in some cases have contributed to the construction of the images.

Thanks also to Bernardo Secchi, with whom I have discussed and designed for many years.

The contents of this book were presented during several seminars. For these opportunities I would like to thank:

Antonio Font and UPC Barcelona;

Jerold S. Kayden and GSD Harvard;

Alexis Sornin and CCA;

Niels Andersen and the School of Architecture of Aarhus;

Christian Gilot and the School of Architecture of Louvain-la-Neuve.

I thank Martina Barcelloni Corte for her precious help in coordinating the English version, Oliver Babel and Luca Ortelli for their valuable support.

scheme libro

concept   sezioni

alcune parti
+ ampie
e approfondite

                alcune più strette
                e finali
un campo   altre più larghe

che contiene campi lavorati
più o meno, ma anche
dei meno lavorati può
essere qualcosa (loro silvestre)

This book explores the design of the city and the territory and the capacity of that design to produce knowledge. I approach questions on two separate planes: the first has to do with research using design tools as a specific form of investigation of reality; the second involves the impact of the progressive, though not unavoidable, marginalization of the architect and the urbanist in contemporary society. I believe that these two planes are closely connected, since doubts regarding one can have significant impact on the other, and vice versa (VIGANÒ 2003; 2005a; 2008b).

Marginal explorations

After lengthy study of John Dewey's *Theory of Inquiry* and of the relationship between the construction of rational decisions and the logic of investigation, Donald Schön refers to the architect's work and to his/her modes of knowledge's formation in order to move away from the idea of action as the mere application of a system of pre-set rules. At the start of the 1980s, Schön's book *The Reflective Practitioner* took its cue from the weakening of the legitimacy of the professional's role and the crisis of expert knowledge and of the very idea of expertise that had taken form in the previous decade. These considerations were flanked by a critique of the academic universe's claim to possess the keys to scientific knowledge, whose practice would thus be simply an application. What interested Schön, on the other hand, was the component of reflection contained in the "practical" work of the architect. The architectural studio, the *atelier*, became the prototype of an education towards "reflection-in-action".[1] Unlike Simon (SIMON 1969) and his research on knowledge-based systems, Schön explored the possibility of a new epistemology, attempting to take advantage of the "window that has opened, we know not for how long",[2] in the strict technical rationality that sees practice as problem solving rather than as cognitive activity. To the contrary, according to Schön, the epistemological statute of the project could be clearly identified, asserted, and taken as a reference by other disciplines.

Today the role of the architect and urbanist in contemporary society has inherited all the doubts that formed during the 1960s and 1970s regarding the ability of experts to interpret social expectations. To these doubts, I add the often-voluntary marginalization and detachment of those who, effectively, in pursuit of the idea THE STRUCTURE OF THE BOOK.

of autonomy of the discipline, reject contact with the issues raised by society. Though only in part, this position explains the delay in the acknowledgment and comprehension of the contemporary transformations of the city and the territory. The eyes of many are simply closed.

"Conversation with a situation", appreciation of resistance to change, reconstruction of the problem and of virtual worlds are the key terms of Schön's epistemological hypothesis, which closely affects us.

The project as knowledge producer

The architect and scholar of the city has always been one of the characters in the human comedy, adapting to different scenes and situations, often revising his position with respect to the society and power. We are reminded of the Leopard of Tomasi di Lampedusa, who feels nostalgia for a bygone order yet dreams of exerting an active influence in the society to which he belongs. Even in countries with a great tradition of urban design, such as Holland, the work of architects and urbanists, even the most recent, is swallowed up by the surrounding territory that observes their exercises with nonchalant indifference. The old Leopard has only himself and some elderly servant with which to converse, in an intriguing and intricate labyrinth of texts and drawings that can at best feed an intimate collection, but cannot be swapped and shared with the rest of the society. Despite this, architects and urbanists have the tools and the capacities to approach many of today's important issues, and their themes are at the center of debates on the contemporary condition.

One of these themes is *complexity*; the knowledge of the architect is not, in fact, the product of a process of separation, but the result of the convergence of multiple disciplines. Accused at times of being superficial, architects and urbanists are ready – with *une tête bien faite* – to confront hybrid, complex, inter-poly-trans-disciplinary and interscalar situations, to echo Edgar Morin (1999). The condition of uncertainty, typical, according to many, of the contemporary era, corresponds to a state that architects and urbanists have always known; inevitably the project has to come to terms with time, with interruptions and changes, with multiple actors and subjects that cannot always be crossed, with practices and lifestyles in constant motion. Their effort moves between two extremes: the attempt to bring change into the project, tackling the theme of its flexibility; and, at the opposite extreme, to identify the elements that can cross time, stand up to their modification, orient and guide processes of transformation. In both cases, one must cope with time and with uncertainty.

Finally, it is the aim or objective of the project that is the great question with which every phase of its construction must come to terms. Reflection on the principle of purpose is an integral part of the architect's and urbanist's job. He, or she, is trained to imagine the future, even forced to imagine it: to project means "to cast ahead", to take a position with respect to the future.

Operations

The project – of spaces, cities, territories – is the architect's and urbanist's main tool, thanks to which it is possible to sustain the hypothesis of the capacity of their research to produce knowledge, as I intend to do here. The fundamental reason for their presence on the human stage (the project on different scales), is seen here as a cognitive device, a producer of new knowledge, a tool for investigating a context and adding new materials to existing knowledge. The project is a form of study and research; it reconstructs, contextualizes, and reorganizes reality.

Many different forms of knowledge exist, and defining the project as a producer of knowledge implies accompanying – if not replacing – the modern idea of systematic, autonomous, objective, and formally structured science with another idea of knowledge, represented and transmitted in different ways. Over the last few decades, many scholars have pointed to a different epistemological configuration of knowledge: "less formally structured, less technical and more social, more contingent, more practical" (GIL 1978). Already, in Dewey, the model of scientific investigation is no longer seen as opposite to research in the field of art or morals: multiple investigative procedures can exist in the same structure. In Dewey's model, investigation marks the passage, "the controlled transformation" from an intermediate situation to one "so determinate in its constituent distinctions and relations as to convert the elements of the original situation into a unified whole"[3], one that is the result of the investigation itself, the elaborated experience. In this hypothesis, the project is more than a neutral mediator between subjects. It takes on a critical role amidst actors, subjects, and places.

In this book I will attempt to examine design activity from a cognitive viewpoint, as "design thinking" (ROWE 1987), a fundamental form of thought and investigation[4]. What I have set out to do is not to reflect on the similarity between investigation and design activity – design is investigation – but to observe the research produced by design activity, recognizing the results in terms of knowledge production of design operations.

The field I have explored does not have to do with the whole set of knowledge produced in the interchange between subjects and contexts in the field of design experience. It is limited to considering the project as a form of research performed by a specific subject (a practical activity oriented towards the transformation of space) that exercises an influence over reality. My purpose is not to bring out disciplinary knowledge or pre-established technique, but to reflect on the kinds of knowledge that are produced during the course of the project activity itself. I am not interested so much in exploring the process by which the thinking of the designer is organized – its procedures, limits and rigidity, its sources and references, its "enabling prejudices"[5] – as I am instead in the knowledge that results from this activity, in certain cases in terms of physical modification of a place, in others as a mental device capable of impacting existing viewpoints and ideas.

Therefore I focus on the architect-urbanist and on that subject's activity, deconstructing certain project operations and gathering them within three main hypotheses, open to further articulations and developments.

Three hypotheses

My first hypothesis is about the project as a producer of knowledge through operations of conceptualization. Here I refer to conceptualization as creation of a space and time of abstraction in which to reformulate thought and gaze, and our imagination concerning contemporary territories. Design activity develops concepts and utilizes and reinterprets concepts from other fields; it steps back from and transforms the contingent; it questions the general value of its own statements, such as when they address "the here and now".
The second hypothesis examines the project as producer of knowledge through operations of description applied in the design activity. The project is a particular form of description, as it unveils situations, constructing a relationship with them; it adapts to them using anomalies, discontinuities, differences; it investigates the presence of underlying structures of order, using repetition and discerning pre-existing rules. The description comes out of the definition, granting individuality to every place. The project describes, and if the description of practices, people, and places becomes its center, a radical theory of transformation of the city and the territory can emerge.
The third hypothesis looks at the project as producer of knowledge through the formulation of sequences of hypotheses about the future. While it is true that, from an etymological standpoint, the project always contains an idea of the future, it is also true that a specific field of research exists that sees the process of construction of the future as a rational, shared sequence of choices. The scenario is understood here as an operation that explores the consequences in space of a hypothetical chain of events, actions, decisions, positioned differently in time. It studies the relationships between different hypotheses of use and transformation of space and their possible coexistence, more than their mutual exclusion.
The knowledge produced by this latter series of operations inserts itself, perhaps even more than in the two previous cases, in a recursive process in which the contribution of the architect and the urbanist permits other actors involved in the transformation of the space to work out new positions. The design process not only enters the various contexts, but also is itself a context and construction of situations in which knowledge is exchanged with others and modified. No abstract knowledge exists, but the result of an experience inside which knowledge takes on its own meaning[6].

The hypotheses that I have outlined move away from the investigation-project analogy: the project is investigation. The three main families of operations rep-

resent three epistemologies, three fundamental places of construction of knowledge that reinforce, if verified, the social role of design as well as the originality
and the necessity of a research that utilizes it.

Geography of a research

This research comes from a long, continuous movement from the project to a
reflection that impacts it and vice versa, gradually accumulating thoughts and
references. At the conclusion of this work, more that at its beginning, I have
found certain confirmations of what, over time, I had ordered inside a series of
notes. I also detailed positions that as a whole form a "reflective" geography of
design thinking. Taking the last twenty years as the time span of reference, at
one extremity we have *Design Thinking* by Peter G. Rowe (1987), which clearly
lays out the positions historically linked to "problem solving". To touch on the salient points: at the start of the 1900s, Associationism describes human learning
as the permanent association of impressions repeatedly received by the senses,
and is thrown into crisis by the school of Würzburg, which proposes the concept
of *Aufgabe*, the "task", the thought endowed with a direction. During the first
decades of the 1900s, the Gestalt movement champions holistic thought that
oversees the formation and organization of information. Behaviorism attempts
to explain human comportment solely in physical — not mental — terms, as a
reaction to stimuli arriving from outside. Particularly in the United States, it has a
lasting influence, with repercussions in the field of reflection on design thinking.
The project as a creative process of problem solving is broken down into a series
of activities (analysis, synthesis, assessment, etc.), furthermore reprising the
tradition of *Beaux-Arts* education, ordered according to the same structure. Rowe
assigns particular importance to Newell, Shaw, and Simon (ROWE 1987: 51) who
introduce information processing theory[7], the "dominant school of thought" (ROWE
1987:55), and to the positions of Christopher Alexander. From these authors, Rowe
takes up the idea of "heuristic reasoning" with which he defines five classes of
research procedures capable of structuring the problem (ROWE 1987: 80): anthro-
pometric and literal analogies; environmental relations; typologies, which anchor
the legitimacy of choice in the past (VIDLER 1977); and, finally, formal languages.
Rowe never completely departs from the "information-processing paradigm" of
the theory of problem solving, though he does point to its devastating crisis, the
result of the attempt to pass from a description of the process to the normative
field of the design of the process, a move that turns out to be an utter failure.
Though only in an analytical sense, Rowe confronts the normative terrain, linked
to values and aspirations; he indicates the role of drawing as a medium that helps
define the design process and proposes a conceptual framework that recognizes
different positions, tied to the modes of production of the object, to the architectural devices utilized, or to the orientation assumed (ROWE 1987: 123). The recognized positions[8] move the action along a specific direction and within two fields

of investigation; the first defines itself in relation to the outside world, while the second is fully inside a disciplinary sphere, inside an idea of the autonomy of architecture and its constituent elements.

The years in which Rowe writes are downstream from the recognition of the crisis of expert knowledge and of the project based on a technocratic and quantitative approach; right in the midst of the return to an idea of architectural and urban design separated from other fields, from the social sciences, for example, and partially – or even fully – uprooted from reality. Rowe invokes a possible coexistence of two "realms of inquiry" (Rowe 1987: 201) and identifies a shared problem: the problem that arises from the need to rediscover a social meaning for design activity, together with the possibility of widening the comprehension of its mechanisms of ideation and construction beyond the narrow circles of sector professionals. His conclusions constitute the starting point of my reasoning, which does not set out to describe – and certainly not to regulate – the cognitive process, but rather attempts to observe the results of certain design operations and to reflect on the knowledge they produce.

Almost twenty years later, the theme is taken up again by Snodgrass and Coyne in the elegantly subtitled *Interpretation in Architecture: Design as a Way of Thinking* (2006). The shift in hermeneutic direction already announced in Schön's text is decisive and unilateral: design activity finds its strongest metaphor in conversation, in the dialogic exchange between subject and object, a situation in which the borderlines and distances between the two entities are lost. The critique of positivist knowledge, relating the design process to linguistic construction and formal logic, leaves no room for appeal. The "hermeneutic circle" makes it possible to escape the logical contradiction caused by the impossibility of comprehending a text when the individual parts or words are read or pronounced, and from their different apprehension in the light of the overall context, which simply explains that formal logic cannot resolve the process of comprehension. The reference is to Gadamer, who defines the process of comprehension in terms of an act of projection, the projection of an overall meaning of the text, while the individual parts or words slowly unfold. The interpretation comes from prejudices that give rise not to an unproductive circuit, but to the acceptance of the finite nature of the process of comprehension and learning of the world, challenged each time anew by the situation in which the project is located, posing questions to which places and persons respond with other questions.

The rejection of the possibility of reconciling the two approaches – logical and dialogic – of any deductive or inductive approach and the total closure with respect to the epistemological component of the design process in favor of the sole hermeneutic component[9] can only be understood, in their vehemence, in the attempt to mark, without ambiguity, the taking of a distance from different and still prevalent opinions. It serves to emphasize that "logic-based models are powerless" (SNODGRASS and COYNE 2006: 51) and to assert the capacity of design activity to prompt thought, assigning it a critical position with respect to the production chain in which architecture and urbanism are inserted.

The hermeneutics of Gadamer is also the reference point, even more recently, of *Creating Knowledge: Innovation Strategies for Designing Urban Landscapes* (VON SEGGERN, WERNER, GROSSE-BÄCHLE 2008). The act of comprehension is understood here as a creative act; it defines a single simultaneous process in which new ideas converge together with the formation of the interpretation. The design of space presents itself as a possible bridge linking different disciplines, the social sciences, and planning. Interpretation requires varied readings that are linked to artistic practices; the sciences of the mind reinforce the hypothesis of a strong link between interpretation and creativity.

With respect to Schön and Rowe, the historical moment has changed greatly. Snodgrass and Coyne observe not so much the crisis of expert knowledge as the weakening of its role, with limits placed on research outside the established paths, not only in academic circles. Policies based on productivist rationality "assume that the role of the professional in the society is clear and not problematic" (SNOD-GRASS and COYNE 2006: 89); the professionals' task, in the field of the construction industry, for example, is to achieve "maximum value", and it is but a very short step from here to their reduction to being part of a mechanism.

What is interesting about the positions of Snodgrass and Coyne, which moreover are very different from those proposed herein, is the attempt to approach the question of the "other", the "fusion of horizons" as Gadamer puts it, widening the hypothesis of the project as a tool for interpreting in an intercultural context, not limited by Eurocentrism (SNODGRASS and COYNE 2006: 258).

A book

My exploration of the epistemological statute of design starts from these reflec-tions, addressing its capacity to produce knowledge and be a device of reality com-prehension, the basis of the construction of a critical vision of the world. The three hypotheses organize the book in three parts, which should not be interpreted as a simplification of design activity. The physical and conceptual "territories" that cross the book are the result of direct explorations, connected with a design prac-tice that has produced original knowledge, as the book asserts. The project neither institutes obligatory sequences nor reduces itself to them, but contains different moves and often uses them simultaneously. The attempt is to give them room, to observe them in sufficient depth to be able to understand their reciprocal neces-sity and presence.

Seen from a certain distance, the crisis addressed by Schön, also marked by the economic recession and the failures of many design efforts, and the simplifica-tion and reduction of the urbanist's and architect's roles seem like two sides of the same coin. Both crisis and simplification, though in different ways, marginalize design thinking and put it into a subordinate position in which it becomes more difficult not only to communicate a critical vision of the world but also to develop it.

[1] On reflection linked to a process driven by rationalization, see BECK, GIDDENS, LASH 1994 and ASCHER 2001: 24-28.

[2] SCHÖN 1993, *Il professionista riflessivo*, introduction to the Italian edition of SCHÖN 1983, *The Reflective Practicioner*, p. 24. On the same pages, Schön talks about a "battle of epistemologies" in progress and the need to reinforce reflective practice.

[3] DEWEY 1939 (trad. it. 1949: 157).

[4] Borrowing and extending to design the expression of Patrick Geddes who used graphs and grid diagrams to reflect on possible innovative relationships between the various terms of his research projects. A support for thought, the "thinking machines" are visual organizations that take stock of the mode of construction of a logical and cognitive process. See also the first part of this book.

[5] P. Rowe uses the terms in reference to Gadamer who, in *Wahrheit und methode* (1960), speaks of *vorurteil* (ROWE 1987: 37).

[6] As in Dewey's model of investigation, "knowledge takes on meaning precisely in relation to the context, the situation. More precisely, in relation to that complex whole that connects reality, the subject and the action in an entwined whole that Dewey defines as experience" (BIANCHETTI 1987); see also: BIANCHETTI 1989.

[7] Newell, A, Shaw, J.C., Simon, H.A., "Elements of a Theory of Human Problem-Solving", *Psychological Review,* n° 65, 1958, pp. 151-166.

[8] They are the functionalist, the populist, the conventionist and the formalist.

[9] "Hermeneutic projection in the design process has nothing to do with the formation of a scientific hypothesis, nor is dialogical questioning in any way akin to the processes of verification or falsification of a hypothesis" (SNODGRASS and COYNE 2006: 51).

CONCEPTUAL TERRITORIES
Part I

The main hypothesis that runs throughout the book is that design produces knowledge through certain operations; conceptualization, investigated in this part, is one of them. The project formulates and develops concepts, and of this ability to utilize, manipulate, and generate concepts I underline the necessity today.

Are epistemological ruptures introduced by the concept or by practices? Bernard Huet, in his last lecture at the School of Architecture of Paris Belleville (HUET 2003), insisted on the founding and disruptive value of the concept even when the practice, for example the new practice of architecture of Brunelleschi, comes prior to the formulation of its concept, in this case on the part of Alberti. The concept, Huet reminds us, is retroactive and transforms.

Does the design of the city and the territory produce concepts? And what operations can we perform with a concept? Is conceptualization a research tool? So the project, through conceptualization, produces knowledge? What are the places of conceptualization of the project? There are many questions that can only be answered right now in introductory, provisional ways.

The first part of the book is structured in three chapters that are very different from one another, and have a different weight: the first has an introductory role and starts with certain reflections on the need to "rethink urbanism" by moving beyond the concepts developed during the construction of the modern industrial city. The hypothesis advanced is that the breakdown of many of the theories of modern town planning requires a new, intense imaginative and conceptual effort. The effort is also that of clarifying the main terms used in this part of the research.

The second and third chapters, "the concepts of the urbanists (1 and 2)", constitute the central and more extensive crux of this part; they select certain concepts represented in the form of diagrams, some of which are very well known, others less so, of urbanists, architects, and scholars of the city. The diagrams are interpreted here as a sort of "compressed file", a closed box that contains the codes for opening and unpacking it, objects to be deconstructed and interpreted.

The conclusions, finally, briefly return to the "concept" and its characteristics. They reflect on the initial and temporary results of research that could continue over a much longer span.

INTERPRETATIVE REDESIGN OF THE DIAGRAM SCHEMA OF THE PLAN OF A SETTLEMENT, TAKEN FROM GLOEDEN, E. 1923, *INFLATION DER GROSSTÄDTE.*

In this framework, the text *Design in Architecture* by Geoffrey Broadbent (BROAD-BENT 1973)[1] represents an important, though controversial, point of reference. It is placed in a moment of renewed faith in the possible role of the architect, which had been weakened by two centuries of separation and fragmentation of knowledge during which more specialized figures, such as the engineer, were often and more frequently found at the center of urban transformations. Paradoxically, the computer, by simplifying the discussion of all the quantitative aspects, can be given credit for having rehabilitated "the architect or, at least, the architectural mode of thinking" (BROADBENT 1988: XV)[2] that translates into the capacity not so much to calculate the structure of a bridge as to answer the question of how one crosses a river[3], to reconstruct the problem by coming to terms with its strategic nodes.

Broadbent reviews the various interpretations of the work of the designer made by scholars of the different types of creativity, and identifies four forms of "design": pragmatic, iconic (when the reference is to a "fixed mental image of what a building should be like") (BROADBENT 1988: 30), analogic and canonic (regulated by systems of proportions, modules, etc.). These forms interact in concrete experience with different degrees of intensity. The techniques of design and optimization of solutions for complex problems, such as System Analysis and Operational Research (O.R.) that still meet with great success at the start of the 1970s, do not challenge the different ways of constructing the project, but simply add new materials, analogies, and canons. They provide tools with which to counter the increasingly harsh criticism of the modern living space and reveal a deeper cultural change, exemplified by the trajectories of the philosophy of science in the same period and in the emphasis placed on systems and processes (KELLER 2005; 2007).

I mention Broadbent's text and those that came after it because they raise an important question regarding the relationship between different forms of knowledge: design, on the one hand, and the human and hard sciences, on the other. The author's conviction, in spite of the faith in design thinking, seems to be that reflection through design feeds on the ideas and concepts supplied by the other sciences. The basis is philosophical, the project an application of concepts proposed in other contexts; projects are "Philosophies into practice" (BROADBENT 1990: 87). An almost idealistic vision permeates the writings and the structure of the books of this influential author; the discussion of the philosophical contributions often comes prior to analysis of design activity.

Things appear to be more complicated. The idea of a project that puts philosophy into practice, that passively feeds on the fount of knowledge to eventually generate sectorial knowledge (in the best of cases a good building, a good plan) is not very satisfying. It fails to clarify one fundamental point: that ideas, wherever they come from, are subjected to transformation and manipulation through design. While in some cases this activity can forecast or produce concepts that later pass into other fields of knowledge (just consider the analogical or metaphorical use different sciences make of spatial concepts, or concepts that reference the forms of the city), more frequently what happens is a reformulation of the initial concept during the course of its translation (and not application) in design. It intersects with others and gives rise to an original product that, in turn, then crosses different fields of knowledge.

In this book I concentrate most closely on the design of the city and the territory; on a number of occasions, the concepts set in motion regarding their design have served to work out more general hypotheses not just about society, but also flow systems, ecosystems, economics, history. In certain cases, they have revealed crucial paths of research for other disciplines. In a perspective of formation of non-monolithic, non-separate zones of expertise, this seems to be one of the fundamental contributions of design as a producer of knowledge.

Two territories cross the first part (Territory 1, "The Project of Isotropy"; Territory 2, "Conceptual Shifts") in which conceptual operations and shifts are the central focus of the reasoning.

INTERPRETATION OF THE
DIAGRAM
SCHEMA OF THE PLAN OF A
SETTLEMENT,
PLATE 2, TAKEN FROM GLOEDEN,
E. 1923, INFLATION DER
GROSSTÄDTE.

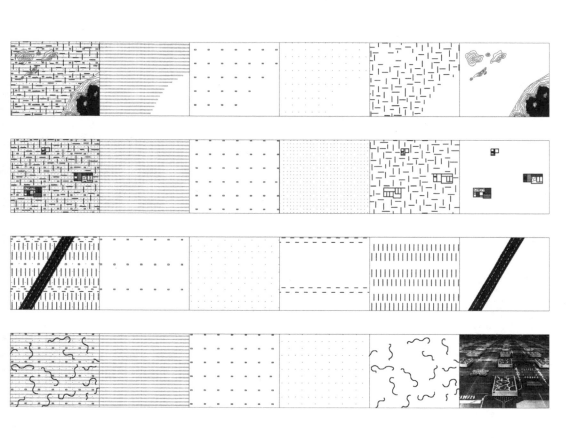

In past years[4], with the terms *descriptive, representative,* or *demonstrative project*, I have attempted to indicate three main families of positions, whose presence I could recognize inside a wide-ranging debate (VIGANÒ 1994). These families made reference to three fundamental project dimensions, three unavoidable and omnipresent dimensions whose intensity, however, can vary remarkably in different moments, and from project to project: the descriptive, representative and demonstrative dimensions.

At the start of the 1990s, the debate was mostly skewed towards the first two. The projects in which the descriptive dimension seemed to prevail over the others were characterized by the return and continuation of a contextual position that focused not only on the form of the territory and its stratification, but also on the new forms of contemporary existence. The attempt to define new images and models of reference more pertinent to contemporary space was contained in the projects with a strong representative dimension. In both families – descriptive and representative – the contemporary city, with its constituent materials, structural and symbolic characteristics, began to present itself as the main object of design research.

In the first family that, for the sake of brevity, I am calling *descriptive project*, I put the research that set out to shed light on the aspects of minimal rationality of contemporary space by describing them; these projects constituted a reaction against the prevailing interpretation of the territory as chaotic, without rules. In the second family, *representative project*, I put projects that attempted to represent these characteristics and transpose them into a model (the *Patchwork Metropolis* of Neutelings, for example, or the praise of congestion of Koolhaas). The enthusiasm and wonder that accompanied the discovery of the contemporary city in those years pushed aside the third dimension, that of *demonstration*, to the edge of the research.

What I intend to assert is precisely the resurfacing of the design's demonstrative dimension and therefore the return to the foreground and the advance of the need to reconceptualize the design of the city and the territory since the last major demonstrative project of modern urbanism. Nevertheless, the social, cultural and economic context within which the project can hope to position itself, and which it can contribute to define, is profoundly different even from that of the recent past.

REDRAWING OF NO-STOP CITY, 1969-1972, ARCHIZOOM ASSOCIATI.

Conceptual breakdowns

In the mid-1980s, the thinking in the disciplinary field of urbanism rotated around the idea that design tools had to be rethought – on different scales. The question of a new form of plan, the relationship between plan and project, fed a long, laborious discussion, producing innovations, reinterpretations, and certain exemplary cases that attempted to re-establish the practice of urbanism[5]. In the meantime something had changed in the city, but not until the 1990s did the focus shift from the renewal of the practices to the identification and construction of a new object of research: the contemporary city.

At first its space seems chaotic, lacking in comprehensible relations, entirely in need of being revealed and redefined. It becomes urgent to develop new keys of interpretation, updated conceptual tools. The definitive exit from the city of industrial modernity, as the background for the slow emergence of a shared level of performance, forced rethinking of the project of Western urbanism – a project that had become implicit[6] – often pursued in an acritical, unconscious way. Many of the concepts behind the modern city seemed to dissolve, touching on materials of a different nature or becoming more radical at the same time. This process has interested all the fundamental categories of the industrial society in which and for which that project had taken form. The concepts of linear progress, of a society based on class, gender, lifestyle, the role of the family (the radicalization of the social role of the family in the economics of widespread small businesses, for example) enter into conflict with an increasing thrust towards individualization. I can recognize at least three major areas of conceptual breakdown.

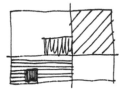

The functional breakdown, or the uncertain margin.   In the contemporary territory zoning, the idea of functional separation has adapted to forceful hybrids and complexities. The profoundly modern conceptualization of the city by zones (MANCUSO 1978), which was initially separated from a social standpoint and then from a functional standpoint, is superimposed on an evolutionary interpretation of urban space, becoming a composite. The city as organism, or as accomplished artwork, which has been completed, is being completed with, to be expanded. The completeness of the city is taken as an objective to be achieved in time and with time. The critique of zoning is widespread and shared in these decades, and in Italy since the 1950s the effort has emerged to introduce different conceptualizations, closer to real processes. The writings of Giancarlo De Carlo in those years are among the most vigorous examples and interpretations (DE CARLO 1964).

In many more recent projects, the attempt to get beyond the great metaphor of the "functional city", the city as a machine composed of separate parts with different connotations, has moved in a few main directions. The first direction overlays areas marked by different functional programs; it theorizes the possible coexistence of functionally and formally heterogeneous parts within complex portions of the city (as in the plan of the airport city in Seoul by Rem Koolhaas, and the recently completed project of OMA for the center of Almere). The move is that of the overlaying of distinct functions, in some cases in a literal, formulaic way, as in the "stacked urbanism" of MVRDV, in others through the juxtaposition of multiple programs in the same place, with the hypothesis that many simple functions (a reminder of the pure functions of which Le Corbusier spoke in the *Ville Radieuse*) can share the same space.

The second direction of research counters the zone with the situation, defining it in terms of spatial configuration identified by a specific pattern. Upon close observation, the Patchwork Metropolis of Willem Jan Neutelings (Neutelings 1990)[7] is composed of various situations, each of which is constructed up against a specific space: for example, living in the woods along a highway, or living around a lake on a golf course. The intentional use of paradox, of unusual juxtaposition, or of the surreal situation, plays the rhetorical role of demonstrating that inhabitable space can be very different from what is communicated through the concept of the "residential zone". In this second case, the close relationship between construction of the landscape and inhabitable space also seems fundamental: the two processes of modification happen in the same moment and enhance each other, producing a new condition.

The third direction of research replaces the logic of functional separation with the hypothesis that the design of the city can be defined by overlapping different systems, each of which responds to different logics and criteria of functioning, as in the plan of Bergamo by Bernardo Secchi and in some of the experiences we have conducted together after that. Each system contains specific rules of settlement and performance in relation to a variegated set of structural models. There are network structures, or sponge structures, when continuity is inevitable or necessary (as in the case of environmental systems or those of mobility); and topological structures, for example in the case of a system of central places where a structure of relations is suggested that is not necessarily based on continuity or on spatial contiguity. The different functional programs can mix and combine in relation to the performance expected of each place, and this is inevitably connected with reflections on the role and identity of each part of the city and the territory.

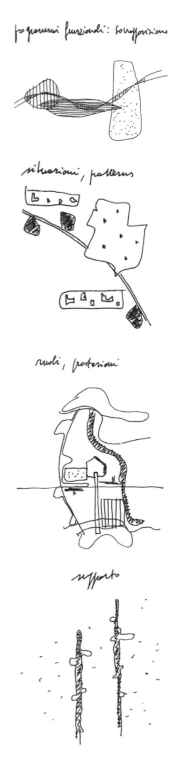

The fourth direction of research returns to the positions of architectural structuralism and distinguishes between a support, composed of infrastructures and
equipment, and the rest of the territory, left up to spontaneous initiatives or ones
that are coordinated in a looser way[8]. The presence of the support, the object of
design, and the opportunity for a more detailed project guarantees the coherence
of a whole in which functional juxtapositions that have not been premeditated can
emerge.

These different research directions, separated for the sake of clarity in the outline
above, but much more intertwined and intersecting in the design experience of
recent years, all call for investigation on the theme of complexity, functional mixing, and overlapping in a territory where functional planning is no longer capable
of conveying an image of modes of use and urban rhythms free of simplification.
They also suggest reflection on the idea of the part and its borders, on the blurring
of urban functional geographies that almost never have established perimeters,
and on their uncertain boundaries. Studying the Japanese city, Vladimir Krstic
(KRSTIC 1997) concentrates precisely on situations of impermanence and change
that have never overly troubled the oriental city, and to which he attributes the,
at least apparently, chaotic character of Japanese contemporary urban space.
What I want to point out here is that the theme of the uncertain boundary not
confronted by modern urbanism is, instead, a shared feature of all the directions
of research examined above[9].

*The breakdown of the concept of density and the ways in which it was proposed
by modern urbanism.*   The contemporary territory, crowded with inhabited
places, is the simultaneous presence of very different and extreme habitat densities and usage intensities. It is characterized by congestion and rarefaction. The
distance between persons, gazes, and objects established in building regulations
and planning norms does not seem to correspond to shared necessities. Themes
of ventilation, light, and introspection, within the profound changes of lifestyles
and material conditions in which they take place, now permit resumption of new
explorations.

Approaching the theme of density, also in the sense of intensity of use of an
urban asset, allows us to reconsider the degree of infrastructuring of the city (of
high technological and infrastructural density; of low technological and infrastructural density; a city technologically very advanced or not very advanced).
In the "Light Urbanism" of MVRDV (1997), urban planning is seen as a discipline that deals not only with the permanent or the lasting but also with the
project of light urbanization that clearly stands out from the monumental parts
and those that are more stable over time. The hypothesis is that if low density
corresponds to low infrastructuring (few paved roads, few networks of distribution and disposal, and so on), then dispersion might be more economical and
sustainable than the concentrated, high-tech city. This hypothesis demands
critical reflection on certain positions and directions of research supported also

in institutional modes, not just in Europe, that take high density as a guarantee of settlement sustainability. To reprise Unwin's motto as "still nothing gained from overcrowding" suggests more caution and awareness of the ambiguity of a question around which judgments that transcend it are often formulated[10].

The breakdown of the materials that make up the city.   In the modern city that fades into the contemporary city, the urban elements also undergo dissolving or changes of state. Functional breakdown and hybridization, the extinguishing of the paradigm "form follows function", the uncoupling of the relationship between the two terms, have contributed to generate new kinds of research on space (ELLIN 1996). The design theme in which this is represented with greatest clarity is the design of the ground, of its thickness, but also of its transformation into something else, infrastructure, ecology, landscape – buildings as infrastructures and infrastructures as buildings, cities as land and landscape as city. The project of the "normal city" (SECCHI 2008; LOMBARDO 2008) that seems to regret the loss of a spatial order produced by a social and economic configuration belonging to the past opposes and reacts to this breakdown.

The changing of margins, centers, and distances in the territory imposes renewed reflection on the design of infrastructures, always seen as a highly hierarchical project. Reflections on the concepts of percolation, density and porosity of the city lead to research that gets away from an *urbanisme de l'axe*, from hierarchy as the sole possible form of organization, and enters a field of horizontal and diffused relations; the reference is to different ideas of modernity (BRANZI 2007; VIGANÒ 2001; 2008a).

So far we have seen the conceptual breakdown of modern urbanism in the contemporary territory. Getting beyond its models implies a renewed effort of measurement of forms and formalization of quantities, of conceptualization and abstraction; it implies a conceptual nomadism (STENGERS 1987) that utilizes theories that are also distant from each other in the attempt to grasp certain general, and not only specific, aspects of contemporary space. The description of the territory tends to transform the observed phenomenon into a model. The concept and the diagram are among the tools of this design research that starts with careful observation of the fundamental characteristics of contemporary space and are the expression of a forceful return to a "structuralist activity"[11], to the elaboration of "guiding models", of tools suited to the situation.

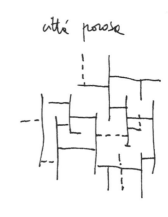

Conceptual tools

I will briefly examine some of the terms that frequently appear in this part of the book: concept, image, schema, analogy, and diagram. This is a useful vocabulary for study in greater depth of the tools of conceptualization covered in the subsequent chapters in order to return, in the conclusions, to the meaning and role of the concept.

Concept.   The *concept*, the project moment and form of diagrammatic representation endowed with its own autonomy, has been widely used in recent years as an instrument of analysis and conception of a project of complexity, taken as an integral part of the contemporary condition. In other words, if today the epistemological task of the project is to hold together what modernity had separated, it is necessary to make a conceptual effort that permits us to elaborate new coexistences. Observing the rise and fall of the production of diagrams in modern urban planning, we notice their frequency in the founding period of the construction of the industrial city, between the mid-1800s and the early decades of the 1900s, and then a resurgence in the design production of the 1950s and 1960s. This is followed, with a very few exceptions, by a drought towards the end of the 1970s and throughout the 1980s, and then a comeback during the course of the 1990s. Donald Davidson writes that conceptual schemes "are points of view from which individuals, cultures, or periods survey the passing scene"[12]: the reference to surveying sheds light on the critical potential of the concept, as a tool of investigation, interpretation, and transformation of reality, but also its belonging to history, to different traditions and paradigms, different worlds in which scholars move[13].

The Italian term *concetto* has been rethought in modern times in the light of Kantian philosophy (*Begriff*), which points to its contribution of generalization and abstraction. This reformulation obscures a more ancient semantic history, very close to artistic practice, marked by the Renaissance and already visible in Dante's polysemic use of the term: "idea, concept, thought, image, intention (in the sense of an intellectual and artistic project), act of the creating imagination"[14]. Linked in the 16th century to the production of representations and *schemata*, in the term *concetto*, the Latin root of *conceiving*, imagining is reinforced. For Vasari the term "denotes a particularly active intellectual art, a conception whose function is to promote the art of drawing as a form of thought"[15]. In the years to follow, the term *disegno* almost covers over that of *concetto* which, in turn, detaches itself from the idea in the Platonic sense and becomes "the intellectual act of a creative freedom exercised on signs, forms, representations"[16], including ingenious ideas, clever inventions, and representations. Imaginative productivity, aesthetic invention, and figurative content are at the center of the *concetto*.

In his reformulation of the German philosophical dictionary, Kant inserts the term *Begriff* and, above all, *begreifen*, the act of comprehending, grasping, in the classification of types of knowledge. *Begreifen* is placed at a higher rank than conceiving, because we can also conceive of that which we do not comprehend, while *begreifen* is a mode of knowledge that utilizes an intuition *per apprehensionem*, as a means to know, to understand. Hegel, rigidifying the concept in a solitary loss of the plural, reinforces it as a figure of knowledge.

In the philosophical world, since the end of the 20th century, we have seen a massive return to the concept and to conceptualization, putting a term back into use – *concept* – which seemed to have lost pertinent philosophical meaning in English-speaking circles (PEACOCKE 1992: 1). This return also has to do with the design of the city and the territory, so it clearly concerns us.

Image.   The role of images, alongside concepts, in the production of knowledge has been an open question for centuries. Descartes assigns the image a materiality that excludes it from consciousness; the image is an object among external objects. It is the borderline of exteriority[17]. So too for Spinoza, who keeps it separate from the theory of knowledge and sees it in reference to the human body; *hazard, contiguity and habit* connect images and memories. Despite of the risk of false ideas, the image is not entirely other than the idea; it is simply a confused idea, and a passage from one plane to another – from cognitive to imaginative and vice versa – is possible. Not only can a passage be formed between image and thought; in the reasoning of Leibniz, there exists a possible continuity between the two different modes of knowledge. The image is a confused idea, but "the image has the opacity of the infinite, the idea the clarity of the finite, analyzable quantity"[18].

The image has not always had good luck. Distrust of images, the uselessness of the image in the cognitive process, corruption and impediment of thought, thought *versus* image: these are reactions that convey a negative viewpoint, disseminated by psychologists and philosophers and widespread until World War I. The image emerges, instead, from the research of Arnheim (1969), the works of Gyorgy Kepes (1954), or those of Marvin Minsky in the 1970s, as an ordering element, a frame that organizes our cognitive processes. It is a data structure through which we encounter new situations; it is an intermediary, also composed of symbols, between reality and mind (BOYER 2003)[19].

Schema.   A fundamental term in Bergson, the *dynamic schema* permits the passage from one plane to the other and is the synthesis "that contains the rules of its development in images"; it contains "the indications of what must be done to reconstitute images" (SARTRE 1956: 61). It is an analytical tool in which the images of Bergson, still elements to be associated, are defined through operations not just of simple juxtaposition but of fusion. According to Bergson invention, comprehension, and memory initially appear in the form of a schema, which, when filled with images, can be modified by them in a dynamic relationship during the course of which mental effort comes into play. Though without clarifying how the schema manages to establish an order in the disorderly crowd of images, Bergson does, interestingly enough, assign the image the role of the intermediary between two fields, the material and the mental, the abstract and the specific field linked to experience.

The schema as the tool of putting into form (*schemata*, i.e. choreography) of images, itself a medium, meets with success in the early years of the 20th century thanks to Bergson and Bergsonism, which nevertheless return quite closely to Kant's definition of the "transcendental schema"[20]. In this same period the schema also encounters favor in the field of urban design and urbanism. The start or conclusion of a process of conceptualization, the schema is also the framework that organizes and structures information in Gestalt psychology, forcefully oriented with respect to external visual stimuli. The schema contains potential, as Sartre – a fierce critic of Bergsonian positions – also recognizes; it is a principle of unity charged with perceptible material, an image impoverished and reduced to a skeleton, "pure determination of the geometric space that claims to translate ideal relations towards spatial relations" (SARTRE 1956: 67)[21]. This term has a long history and an important position in design conceptualization, though it is almost always utilized as a posteriori synthesis, an attempt at simplification, and not as a tool to construct new relationships between images and concepts. In Greek the *schema* designated not only the form but also the figure, a dense term that belongs to the field of linguistic and literary analysis, more than to the philosophical dictionary. Choreography is composed of forms and figures, and thinking through figures belongs to the design field.

Analogy.   The use of analogies is frequent in design activity. They permit us to give a representation to phenomena and questions we would otherwise not be able to approach. By means of analogies we juxtapose known and unknown things, establishing a comparison between them; or we use them to explain concepts that would otherwise be hard to communicate. This definition of analogy applied by Broadbent, who picks it up from Operational Research, is also useful to define the model, in the sense of analogical representation of the phenomenon being analyzed. The insistence on the incompleteness of any model (it is not the phenomenon, and it represents it only in an incomplete way) may seem pedantic, but it is necessary at the start of a path that crosses several of the concepts proposed by urbanists. At times the loss of distance between the model (bearer of the analogy utilized) and observed reality comes to the surface.

Operational Research proposes a distinction between "iconic", "analogue" and "symbolic" models. While the iconic model has an established tradition in the history of architecture and urban planning, and the symbolic model fundamentally regards fields such as linguistics or mathematics, the most interesting aspects have to do with the analogue model. It transcends the limits of the first model, forced to resemble what it represents, and avoids the constraints of the symbolic model, after the naive attempt of System Analysis to restrict architectural and urban reasoning to a set of equations. Dealing with phenomena and processes, the analogue model in its different forms – drawn, materic, etc. – is a device of conceptual translation. From maps to models to diagrams of the city, the analogue model is an act of imagination that changes the properties of the examined objects and represents them by making use of a wide spectrum of possibilities. The model prompts us to bring out the structure of the object, phenomenon, or question we intend to represent.

Analogy belongs to logic. It uses the resemblance between objects and infers the knowledge of one of them to approach and understand the other; the resemblance must involve relevant aspects, while the difference must regard only what is irrelevant (BROADBENT 1973). The metaphor, a figure of speech, pertains to the transfer of characters or properties from one object to a different object; it is a way of borrowing, of transferring into different contexts. The metaphor is a transaction, the passage of certain elements from one context to another, shedding a different light on the latter.

Diagram.   In recent years, in relation to forceful activity of conceptualization, particularly in the field of architecture, we have seen a return to reflection on the *diagram* as a tool of representation and conception of the project. Without retracing the entire debate[22], its salient points have to do with the interpretation of the diagram as an explanatory device that analyzes, explains, and communicates the project, and of the diagram as a constructive and generative device. In the latter case, the diagram asserts and reinforces the analytical dimension of the project, a tool of investigation, and a possible object of analysis in its own right, which can be parsed as the result of operations of the same nature (MENNA 1975). What interests me here about the diagram, as I will show more thoroughly in the next chapter, is its unquestionable capacity to convey concepts; it constitutes one of the most important "techniques and procedures of architectural knowledge" (CONFURIUS 2000)[23].

The diagram can be applied in different moments of the design process and can be used with different aims, including that of the critical analysis of the project, or to bring out implicit diagrams that have never been made explicit (EISENMAN 1998: 27-29). If the architectural diagram is at times the base on which to trace the definitive lines of the object, in the field of urban and territorial design, the diagram clearly takes on the role of mediation between concepts and images, between figures, analogies, and metaphors coming from heterogeneous and often discordant sources.

The concept (represented in the form of a diagram) is both a form of representation and design thinking, a space of conception in which a condition of freedom and expression exists with respect to established discursive formations, in which to hypothesize relations and dimensions; it is a form of organization of the experience and representation of knowledge (ROWE 1987; BOUDON 2004: 43).

The space of representation plays a fundamental role in the construction of the project, which is allographic art: the necessity, that is, to be represented through the medium of drawing, just as music is represented through the score (Goodman 1968). A change of representation almost always implies a different conceptualization; solving a problem means representing it in a different way (Simon 1969; Boudon 2004: 67).

[1] All references are to the 1988's edition.
[2] "[…] to rehabilitate the architect or, at least, the architectural mode of thinking" (Broadbent 1988: XV).
[3] Broadbent cites the opening talk of the *Conference on Design Methods* (1962) by D.G. Christopherson; See: Jones and Thornley 1963.
[4] This chapter returns to and extends the reflections contained in Viganò 2000a.
[5] In the European and Italian context, for example, the experiences of Oriol Bohigas and Manuel de Solà Morales in Barcelona; in Italy, of Bernardo Secchi at Jesi and Siena.
[6] I began to think about this theme by deconstructing the plan of Giovanni Astengo for Bergamo, a monument of Italian urban planning (Viganò 2000b).
[7] *Patchwork Metropolis*, a study by W.J. Neutelings with W. Sulster, P. van Wesemael and E. Winkler, 1989-1990; also see *El Croquis*, n° 94, 1999: *Neutelings Riedijk 1992-1999*.
[8] The fourth position consists of a review of the structuralist hypotheses of the 1960s, in particular those of the Smithsons and Habraken. In a more recent text, Habraken (1998) approaches certain questions in ways that partially differ from those of the past, emphasizing not so much the rigidity of the support as the opening of the different products to the continuous redesign linked to their use. This does not imply that the reinterpretation of architectural structuralism − especially in housing design but also to some extent in the design of open space − has led to interesting hybrids. "Ground design" ("*progetto di suolo*", Secchi 1986) is also part of this reinterpretation.
[9] Also see the essay by Christopher Alexander, *The City is not a Tree*, mentioned in the pages to follow.
[10] See, among others, *Les Annales de la recherche urbaine*, 1995, n° 67; (Amphoux, Grosjean, Salomon 2001); (Fouchier 1994; 1998); (Heynen, Vanderburgh 2003).
[11] Barthes 1976 (in particular "The Structuralist Activity", [in] *Lettres Nouvelles*, 1963); also see Infussi 1995.
[12] Davidson 2001: 183.
[13] Davidson cites *The Structure of Scientific Revolutions* by Kuhn (1962). World is a Heideggerian term and makes reference to something more vast than culture, to a set of premises in which tradition, place and the quotidian, etc., have a role.
[14] *Vocabulaire européen des philosophies*, Seuil-Le Robert, 2004, pp. 251.
[15] *Ivi*, pp. 252.
[16] *Ivi*, pp. 253.
[17] Sartre 1956: 7.
[18] Sartre 1956: 11 (first edition, 1936): "l'image a l'opacité de l'infini, l'idée, la clarté de la quantité finie et analysable".
[19] In a reflection on cognitive models, Christine M. Boyer (2003) lists the symbolic paradigm, the "connectionist" model, perception and associative models.
[20] "[…] the 'transcendental schema' is an intermediate representation between sensibility and intellect. Which makes possible the application of categories to phenomena, i.e. the elaboration of sensible experience in the light of the pure concepts of the intellect" (entry "Schema", *Enciclopedia Garzanti di filosofia*, 1961).
[21] Sartre references, in particular, the use in mathematics of the term *schema* to indicate the planar or solid figure.

22 The end of the 1990s is marked by many publications on the diagram, in particular in certain magazines. The earliest and most famous include: *ANY* n° 23, 1998: *Diagram Work*; *Oase* n° 44, 1998; Daidalos n° 74, 2000: *Diagrammania*; *Fisuras*, July 2002; *Lotus International* n° 127, 2006.
23 CONFURIUS 2000.

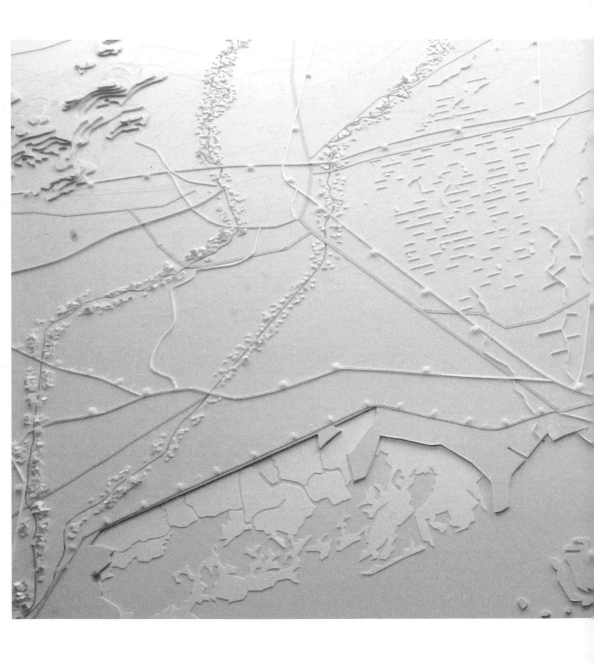

The metropolitan area of Venice is an European territory of diffused settlement. It is marked by an isotropic condition that has informed its processes of transformation across history. The hypothesis behind the research, "Water and Asphalt: The Project of Isotropy", is that these characteristics, now subjected to forceful processes of ranking, represent an opportunity to develop a new territorial vision.

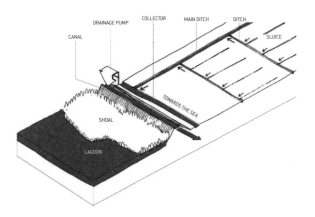

THE ELEMENTS THAT REGULATE
DRAINAGE AND RUN-OFF OF WATER
IN THE RECLAMATION AREA.

B. SECCHI, P. VIGANÒ, *WATER AND ASPHALT: THE PROJECT OF ISOTROPY*, WITH PHD STUDENTS IN URBANISM (M. BALLARIN, M. BRUNELLO, N. DATTOMO, D. DE MATTIA, E. DUSI, V. FERRARIO, S. GIAMETTA, E. GIANNOTTI, M. GRONNING, T. LOMBARDO, P. A. MARCHEVET, MCOISANS, J. (CENTRE DE RECHERCHES SUR L'ESPACE SONORE & L'ENVIRONNEMENT URBAIN-GRENOBLE), M. PATRUNO, M. PERTOLDI, S. PORCARO, C. RENZONI, A. SCARPONI, L. STROSZECK, M. TATTARA, F. VANIN, F. VERONA, G. ZACCARIOTTO, A. ZARAGOZA), IUAV UNIVERSITY OF VENICE – 10TH ARCHITECTURE BIENNALE, VENICE, 2006.

Isotropic representations

The notion of "isotropy" refers to a body, substance, or phenomenon that has the same properties in all directions. Descriptions of the metropolitan territory of Venice, as of many territories of settlement dispersion in Europe, often make use of terms and representations, such as grid, nebula, dispersion, diffusion, etc., that can be connected to isotropy. As a metaphor, isotropy also groups together various generic forms represented both in the physical reality and in the ideal interpretation of the territory.

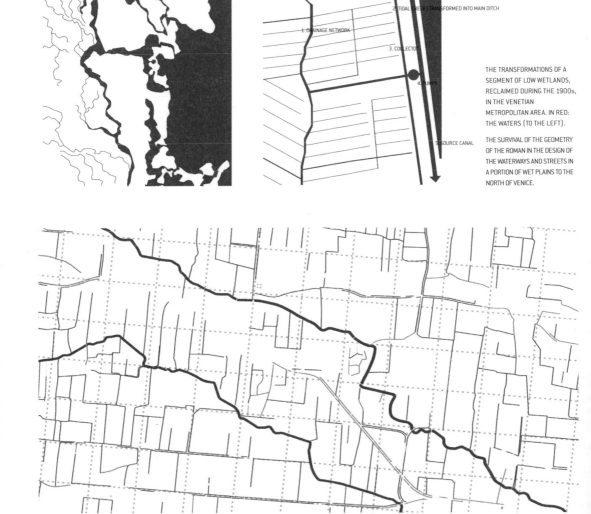

THE TRANSFORMATIONS OF A SEGMENT OF LOW WETLANDS, RECLAIMED DURING THE 1900s, IN THE VENETIAN METROPOLITAN AREA. IN RED: THE WATERS (TO THE LEFT).

THE SURVIVAL OF THE GEOMETRY OF THE ROMAN IN THE DESIGN OF THE WATERWAYS AND STREETS IN A PORTION OF WET PLAINS TO THE NORTH OF VENICE.

A fractal territory and important processes of rationalization

Isotropy is the result of successive processes of territorial rationalization, mainly linked to the regulation of the system of waters.

The notion of "isotropy" addresses all scales: from the smallest, the isotropic network of small canals and the dispersion of isolated single-family houses; to the largest, that of the big infrastructural networks. Nevertheless, recent projects and others in progress, connected above all by the street network, run the risk of greatly reducing movements in the "sponge" of the smaller streets.

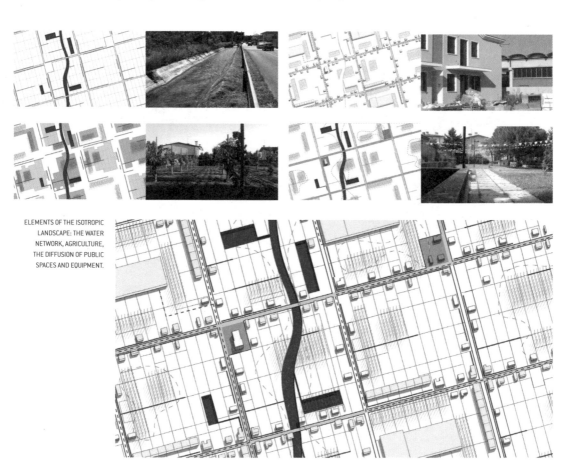

ELEMENTS OF THE ISOTROPIC LANDSCAPE: THE WATER NETWORK, AGRICULTURE, THE DIFFUSION OF PUBLIC SPACES AND EQUIPMENT.

Five paradoxes

The Venetian territories of dispersion are now going through a major mutation from which certain paradoxes are emerging.

The paradox of water: in one of the biggest reserves of water of Europe, the demands of consumers – from domestic use to agriculture – cannot be satisfied, and the territory is subject to frequent flooding.

The paradox of agriculture: the agricultural territory, even today, represents the largest portion, but it runs the risk of depletion and banalization. Deprived of European subsidies, agriculture will have to rethink its role.

The paradox of mixité: the new and recent resistance against function mixing and the limits of a territory polluted by the widespread presence of industrial activities not served by the sewer network.

The paradox of public spaces: the crisis of the traditional concept of urbanness and, at the same time, of the public space of modern welfare.

The paradox of isotropy, finally: a territory that is both diffused and congested, in the midst of progressive hierarchization, in which it is difficult to pursue shared objectives.

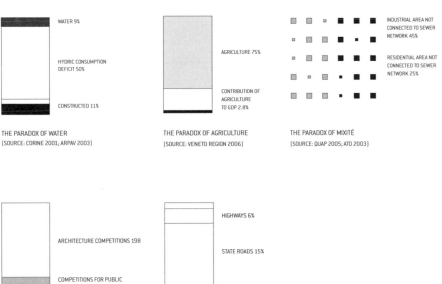

WATER 9%

HYDRIC CONSUMPTION
DEFICIT 50%

CONSTRUCTED 11%

THE PARADOX OF WATER
(SOURCE: CORINE 2001; ARPAV 2003)

AGRICULTURE 75%

CONTRIBUTION OF
AGRICULTURE
TO GDP 2.8%

THE PARADOX OF AGRICULTURE
(SOURCE: VENETO REGION 2006)

INDUSTRIAL AREA NOT
CONNECTED TO SEWER
NETWORK 45%

RESIDENTIAL AREA NOT
CONNECTED TO SEWER
NETWORK 25%

THE PARADOX OF MIXITÉ
(SOURCE: QUAP 2005; ATO 2003)

ARCHITECTURE COMPETITIONS 198

COMPETITIONS FOR PUBLIC
SQUARES 55
COMPETITIONS FOR PARKS 9

THE PARADOX OF PUBLIC SPACES
(SOURCE: EUROPACONCORSI 2006)

HIGHWAYS 6%

STATE ROADS 15%

LOCAL ROADS 12.5%

THE PARADOX OF ISOTROPY
(SOURCE: ISTAT 2000)

INSERTION OF A "TUBE" INSIDE
THE "SPONGE": CUT AND
CROSSINGS RECONNECTION
AND SELECTION HIERARCHICAL
RANKING.

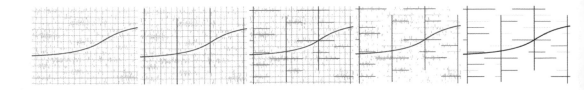

Overturning the paradoxes

The paradoxes reveal a crisis of the idea of territory as infrastructure. They are investigated here by hypothesizing a breaking point and new functional and formal models, starting with the diffused systems of water and asphalt.

Thus the paradoxes of water, agriculture, functional mixture, mobility, and public space can be reversed in new themes in which the isotropy of the Venetian metropolitan territory is no longer just the result of its long history, but becomes a rational figure, a resource, and an inspiration for contemporary projects.

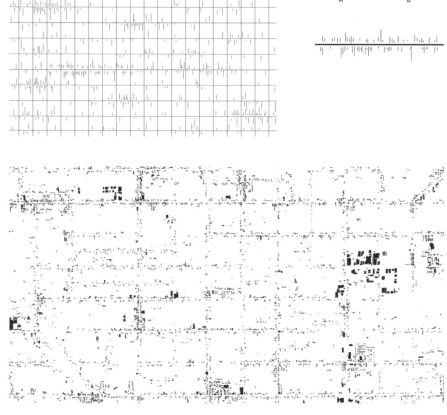

THE SPONGE: AN OSMOTIC
RELATIONSHIP BETWEEN
SETTLEMENT AND ROAD
INFRASTRUCTURE.

New conditions for a project of isotropy

The hypothesis that guides this research is that today conditions exist for a rethinking of the project of isotropy in the Venetian metropolitan region. Its most important elements can be the result of innovative reflection that starts with its main infrastructural supports: the complex network of water and roads.

More space for water.   Reuse of the hundreds of gravel quarries in the dry plains as lamination basins; re-naturalization of rivers and canals. The lagoon is at the center, for both symbolic and ecological reasons: the metropolitan area almost coincides with its "run-off basin".

Minimum 10% of territory for new forests.   In the plains new agroforest systems, productive forests, and urban forests are necessary; new water-saving technologies to irrigate the plains.

Agriculture as a multifunctional landscape: after 2013, agriculture will have to renegotiate its extensive presence in the territory.

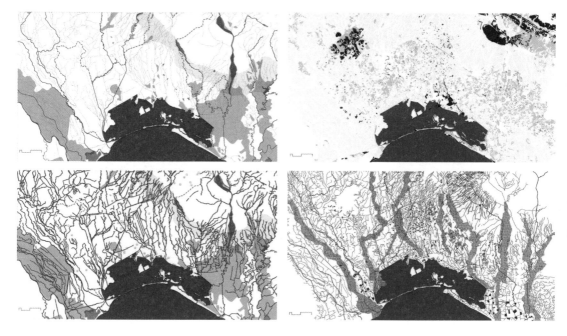

MORE SPACE FOR WATER: CURRENT STATE AND SCENARIO.
IN RED, THE WATERS;
IN GRAY, THE FLOODABLE AREAS;
IN BLACK (DOTS), THE QUARRIES.

MINIMUM 10% OF TERRITORY FOR NEW FORESTS: CURRENT STATE AND SCENARIO.
IN RED, THE WATERS;
IN BLACK (BELOW), THE NEW FORESTS.

Sponges

The sponge, more than the network, is the typical figure of an isotropic area. It is characterized by movements of percolation – of water, people and practices – more than strong hierarchies. The Venetian metropolitan territory is formed by a set of "sponges" with different histories, characters, and landscapes that are the result of rules dictated by rationalities, which should be interpreted. Clarification of the different principles of integration between water, streets, and settlements can generate new rules of construction of the territory.

A mesh of railroads, streetcar lines, and navigable canals intersects the facilities of diffused welfare; to live no more than three kilometers from a railway station; a diffused parking system to leave cars and take bicycles; an isotropic network for slow mobility.

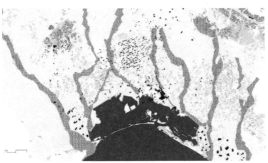

MIN. 10% OF TERRITORY FOR NEW FORESTS: THE NEW
FORESTS (IN BLACK), THE WATER (IN RED),
THE AGRICULTURAL AREAS (IN GREEN AND OCHRE).

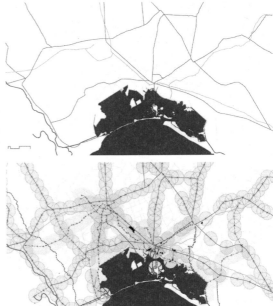

A NEW PUBLIC TRANSPORT NETWORK: CURRENT STATE AND SCENARIO.
IN RED: THE WATER, WITH HATCHING FOR NAVIGABLE ROUTES,
IN BLACK: CONTINUOUS THICK LINE, THE RAILROADS; THIN DOTTED LINE, THE TERRITORIAL
STREETCAR LINES.

Figures of isotropy

Isotropy as ideal figure. In the history of ideas, or the history of art, the figure of isotropy has at times represented an ideal to be pursued. For example, in relation to the utopian conceptions of space of the avant-garde, notions like the continuum or spatial interpenetration can be seen as coming close to an isotropic ideal.

Isotropy as figure of political rationality. In political terms, isotropy is a device of appropriation and control of the territory. Outstanding examples include the *centuriatio*, the colonization of the Venetian territory by the Romans; the re-appropriation of the territory through "reclamations" after the first medieval fragmentation; the new connections between points of centrality. Isotropy, finally, is also a figure connected to the idea of a democratic society.

Isotropy as a figure of economic rationality. In economic terms the question is: up to what point is isotropy capable of efficiently structuring the territory? From this standpoint, it is important to consider the organization of labor and the spaces of everyday life, the technological level and distribution of resources, and the efficiency of the various networks in relation to a territory of settlement dispersion.

Isotropy as a figure of ecological rationality. Even in the regular presence of green patches, the territory of Venice represents a completely manmade habitat. This has significant implications: avoiding confusion between "vegetation" and "nature", this simple fact requires a global rationality that connects all the interventions to a common ecological denominator, without overlooking any individual element or connection. The most important ecological themes are connected with the water that crosses and infiltrates the entire territory.

WATER AND ASPHALT. A PROJECT OF ISOTROPY. MODEL OF THE METROPOLITAN AREA OF VENICE: VIEW FROM THE MOUNTAINS TOWARDS THE LAGOON.

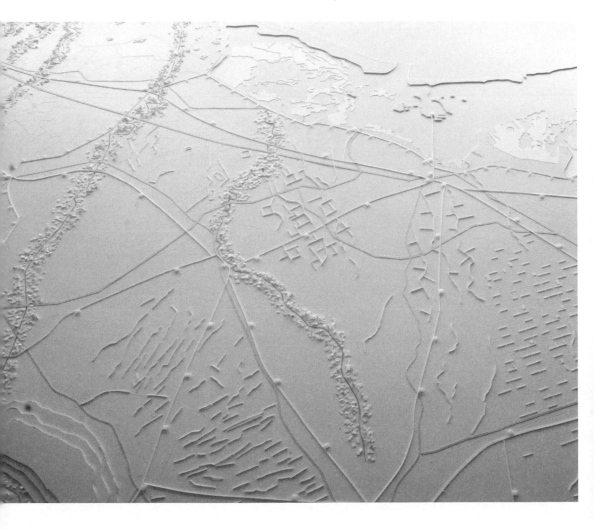

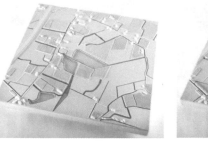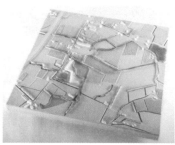

SCENARIO: REUSE OF THE
QUARRY AREAS AS LAMINATION
BASINS: CURRENT SITUATION;
RE-NATURALIZATION OF THE
VARIABLE-SECTION CANALS
AND NEW FOREST STRIPS;
CONSTRUCTION OF A NEW
INHABITABLE SPACE.

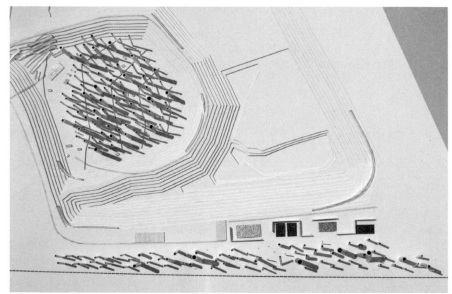

P. VIGANÒ, U. DEGLI UBERTI, G.
LAMBRECHTS, T. LOMBARDO, G.
ZACCARIOTTO, PILOT PROJECT
OF RECOVERY OF CAVA
MEROTTO AT CONEGLIANO WITH
MAINTENANCE OF THE WOODED
WETLAND FORMED IN THE
QUARRY LOCATED IN THE DRY
PLAINS, 2006.

New rules

Rethinking the new rules of construction of the territory starting with the isotropic character of its fundamental relationships.

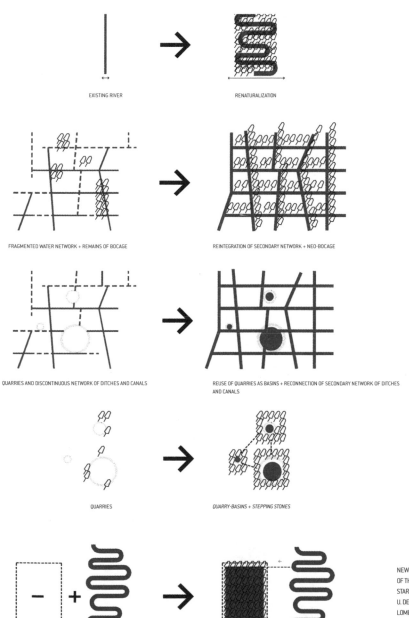

NEW RULES OF CONSTRUCTION OF THE DIFFUSED TERRITORY STARTING WITH WATER. P. VIGANÒ, U. DEGLI UBERTI, G. LAMBRECHTS, T. LOMBARDO, G. ZACCARIOTTO, PILOT PROJECT OF RECOVERY OF CAVA MEROTTO AT CONEGLIANO, 2006. TAKEN FROM *LANDSCAPES OF WATER*, 2010

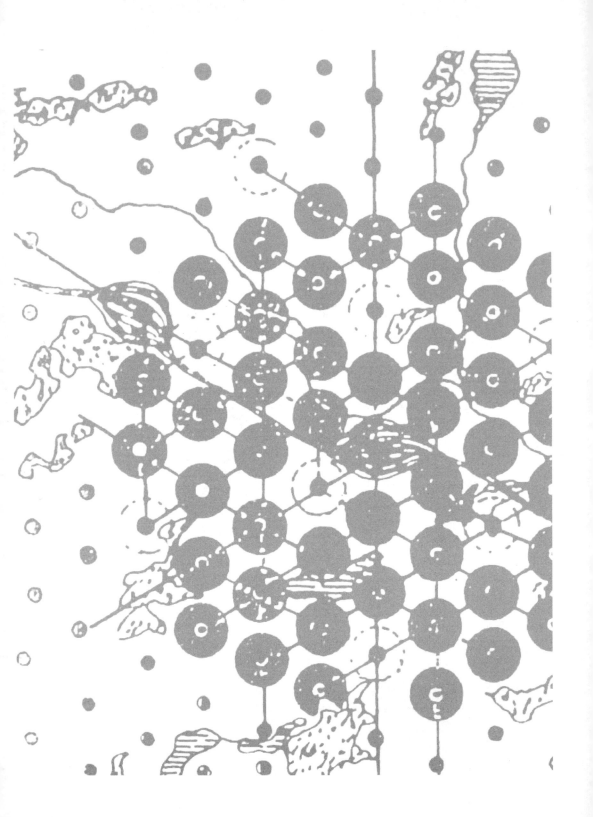

By *concept*, as I tried to demonstrate, I mean the idea and its representation – the conceptualization and graphic expression through a drawing with a diagrammatic character. The interpretation of this particular family of drawings requires, as in any effort of interpretation, a short introduction. I will mention only the two references that have most directly influenced this work: *The Limits of Interpretation*, by Umberto Eco (1990), and *Leçons de musique*, by Pierre Boulez (2005).

From the first, I have gained the conviction of the autonomy of the work (in this case the diagram) with respect to the viewpoint of the author and the various readers. Though in any diagram there exist texts, projects, stories, biographies that to some extent take part in the construction of an interpretation, I have focused on the autonomy of that little, dense work that is the diagram, or the diagrammatic sketch. Following Boulez, I have reinforced the idea that the philological (often "academic", as the author writes) interpretation is not always the most fertile, and that it is worthwhile to take as a whole the creative potential inherent to each interpretation. Precisely during the course of the interpretation, we introduce a conceptual shift that either can reveal unforeseen or simply hidden aspects or can transplant the initial concept on different grounds that transform and reconstitute it.

The work I am doing here is not that of the historian, though the concepts I observe do belong to the history of urbanism. It is more like an interpretation of the work in which room exists for its execution/reproduction. The English verb *to play* and the French verb *jouer*, applied for musical but also theatrical performance, help to clarify this position. It views interpretation/reproduction as an original act (something no one tends to doubt in the case of musical or theatrical interpretation), not just a mechanical reproduction of a score or a text. A rigorous reading, which is only a slow, precise reading through the reproduction of the work – in this case its redrawing – can permit us to enter into the constituent logic of the diagram, granting perfect freedom in its deconstruction and interpretation.

The diagrams that I examine cover a few selected themes and make not attempt to provide an exhaustive outline of the production of concepts over the last century; instead, the selection corresponds to the centrality of certain themes in the design of the city and the territory today. The retraced genealogies are striking for the inertias and continuities they bring to the fore: running through the history of modern urban planning as if it were a large illustrated book[1], we can observe the rise, fall, and settling in projects of certain ideas that introduce new relationships

SETTLEMENT PLAN DIAGRAM, PLATE 1, TAKEN FROM GLOEDEN, E. 1923, INFLATION DER GROSSTÄDTE (DETAIL).

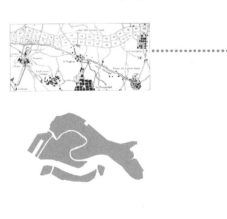

REDESIGN OF THE LINEAR CITY
OF A. SORIA Y MATA: LINEARITY
AS FORM OF DEVELOPMENT
OF SETTLEMENTS ON THE
TERRITORY; THE RELATIONSHIP
WITH THE CENTRAL CITY AND
THE POINT-CITIES.

COMPARISON: VENICE AND THE
LINEAR CITY AROUND MADRID.

between things, that propose original concepts of space. The diagram, in the best-known examples, is a tool of analysis and design of the new urban dimension, the new form of the city, and the relationships it constructs with the rest of the territory.

The diagrams I take into consideration have been selected from the many produced during the last century and towards the end of the 19th century. They explore the transformations of relationships of production and practices of use of space, approaching three distinct groups of themes: the new scale and the new form of the city; the materials of its design and, in particular, the emergence of the background, the "void"; and the project forms, which are the spatial and procedural structures involved in the production of contemporary space.

A new scale, a new form of city

The first diagrams of modern urbanism are not by planners or architects, but by economists like Von Thünen or biologists like Geddes, and come after the diagrams of Enlightenment rationality by physicians or philosophers such as Jeremy Bentham (VIDLER 2000). In a discipline that was just taking form, with blurred boundaries wavering between distant fields of influence, the ring diagram of land use around a city by Von Thünen (1826)[2] or the section technique used by Geddes (1915) are descriptive of an analytical model and a set of ideal relationships.

Soria y Mata, though he had a scientific background, was neither an architect nor an engineer (COLLINS 1968: 16). His diagrams and projects for a linear city (1882) avoid the traditional representation of the city with a center and outskirts, conveying the idea of a networked territory composed of segments of linear city that provide infrastructure, or "vertebrae", for the existing city. With respect to the diagrams of Howard and the German school in this same period, or a few years later, which represent the growth of the urban territory as the gemmation of smaller and decentralized units, as extension of the existing city in decreasing density that is interrupted by green wedges, the linear organization of Soria y Mata introduces a type of city expected to play the role of infrastructure for a wider-ranging territory. The "vertebrate city" gains meaning only because it is not isolated, but is the support of a set of cities: from Cadiz to St. Petersburg, Beijing to Brussels. The rationality of the linear design as opposed to the radiocentric form lies in the logic of modes of infrastructure: streets, railways, gas, water, electricity, and telephone lines are placed along a preferential channel; the facilities are organized in terms of episodes[3].

The linear city is a simple concept, writes Collins, who comments on it extensively. It represents a "natural process" (in quotation marks in the text) that has always been known. Halfway through the 1960s this process seems to have taken on, for the English critic, "a new and terrifying dimension, as the tentacular spokes of adjacent cities gradually intersect, producing those enormous, chaotic belts that represent one of the most serious problems of our era" (COLLINS 1968: 21). We are struck by the shortsighted, contradictory interpretation of the historian, who fails to grasp the absolute coherence between the growth of the city in the 20th century, with its "tentacular spokes", and the idea of the linear city. The linear form, however, does have some advantages: it is open, it does not hamper successive configurations, it permits incremental achievement and has an "abstract, systematic and versatile character" (COLLINS 1968: 35). However, precisely this misunderstanding lies at the origin of the lack of evolution of

HILBERSEIMER (1940)   SORIA Y MATA (1802)   ALEXANDER (1963)

TWO PRINCIPLES OF LINEAR ORGANIZATION: THERE IS NO DISTINCTION BETWEEN INFRASTRUCTURE AND CITY, THE INFRASTRUCTURE IS A SUPPORT OF CLUSTERS.

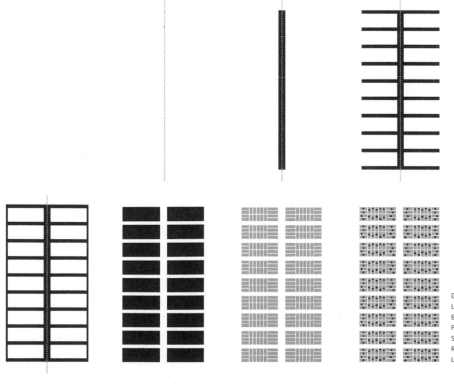

DECONSTRUCTION OF THE LINEAR CITY PROJECT: BAND OF NETWORKS AND PUBLIC TRANSPORT, MAIN STREETS, CROSS STREETS, REARS, URBAN BLOCKS, LOTS, AND EDIFICATION.

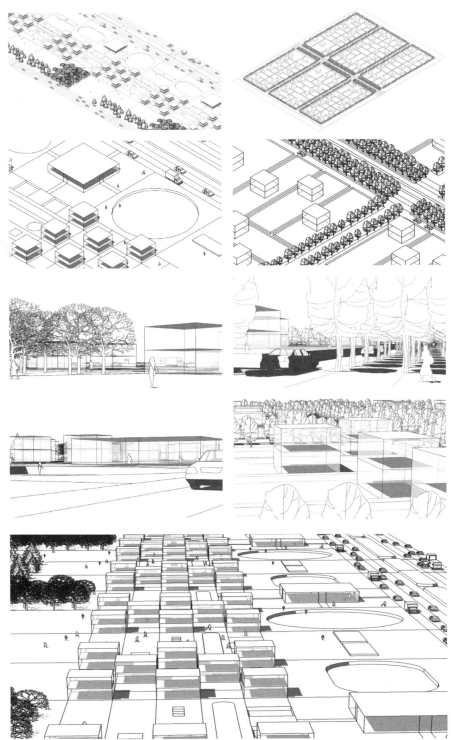

EXAMPLES OF LINEAR CITY
PROJECTS: VARIATIONS IN THE
RELATIONSHIP OF SETTLEMENT
WITH INFRASTRUCTURE, WITH
OPEN SPACE AND DIFFERENT
DEGREES OF PERMEABILITY.
IN THE COLUMNS: BRIGATA OSA,
1920; A. SORIA Y MATA, 1882;
C. ALEXANDER, 1963;
L. HILBERSEIMER, 1940.

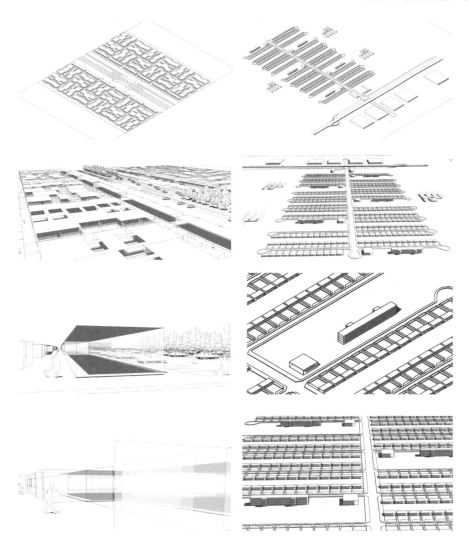

the hypothesis of Soria in contexts different from the initial project of single-family houses and gardens partially completed in Madrid. Certain exceptions, such as the "five-finger plan" of Copenhagen and the more recent Ørestad project, also in Copenhagen, confirm the timeliness of the linear concept.

The radical character of the linear diagram has often been emphasized[4], but less has been said about its ability to structure the new urban territorial scale, with the construction of networks capable of absorbing the existing city inside new spatial and functional concepts: "For new *line cities* that connect today's *point cities*. In this way an immense triangulation would be traced on the surface of the earth…" (SORIA Y MATA 1968: 299). The linear city is imagined as a tool of colonization and resettlement of abandoned and underutilized territories of the entire world.

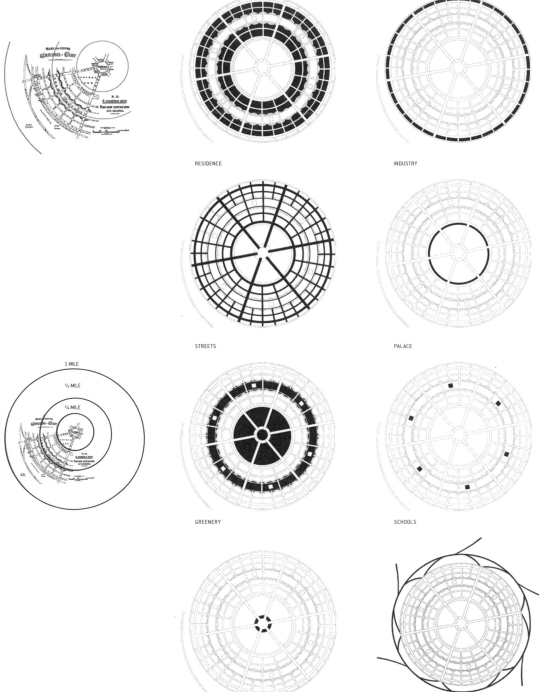

RESIDENCE

INDUSTRY

STREETS

PALACE

1 MILE

½ MILE

¼ MILE

GREENERY

SCHOOLS

DECONSTRUCTION OF THE
DIAGRAM OF THE GARDEN CITY
OF E. HOWARD.

COMMUNITY CENTER (THEATER, LIBRARY, TOWN HALL, HOSPITAL)

RAILROAD, THE LIMIT

In 1898, Howard, in the wake of a rather negative American adventure, accompanies his text *To-morrow: a Peaceful Path to Real Reform*, reissued in 1902 with the title *Garden Cities of To-morrow*, with ideograms that clarify the rules of construction of a *Town-Country*[5]. A summary of ideas and of a debate carefully outlined by Howard, the diagrams put together in a new system of relations familiar urban materials like the boulevard, the avenue, and the crescent, the public park bordered by the Crystal Palace, the cultivated countryside, and the single-family lot. Though it is "a diagram only", as is written beside each one, specifying in the *Garden City Diagram* that "Plan cannot be drawn until site selected", and in the *Ward and Centre* detail that "Plan must depend upon site selected", many of the diagrams are provided with a scale of measurement and therefore illustrate dimensions, surfaces, and distances in the new urban form.

The scale – in this case the new scale of the design of the city, potentially composed of units limited in their expansion – is an integral part of the formulation of the diagram, just as it is an integral part of the conception of a new inhabitable space. The scale of measurement makes it possible to understand the size of the territory involved in the design reflection and the respective dimensions of the objects represented, together with their proportions. This point becomes interesting in relation to the long subsequent debate on the relationships between open and constructed spaces, and on their forms and sizes.

My hypothesis is that the scale of measurement vanishes from diagrams when the dimension of the city, in particular its new, more extensive scale, becomes a given, incontrovertible fact, and no longer a founding or fundamental part of the reconstruction of the relations among things. This interpretation might seem provocative against the backdrop of the wealth of readings and projects on the city-territory that arrived later, in the 1960s, or in even more recent times on the territories of dispersed settlement, but I believe it deserves further assessment. What I mean is that reflection on the scale of the city is pertinent when the city is the opposite of something else, generically indicated with the term "country". Then it is important to measure both and to reveal the ways in which the dimensions of the city increase while those of the country diminish. However, when the city becomes the wider urban context familiar to us today, this continuous measuring and reflection on the different scales becomes irrelevant, and therefore many diagrams lose their reference to measurement.

Patrick Geddes used the diagram as a "thinking machine"[6]. His famous section contained in *Cities in Evolution* and frequently represented is influenced by the image of the valley used by the anarchist geographer Élisée Reclus as a metaphor of the development of civilization. From the social analysis of Frédéric Le Play, Geddes picks up the categories of "Lieu, travail, famille", somewhat loosely translated as "Place, Work, Family". The region represented coincides with the basin of a river, from the sources to the sea[7]. Hilberseimer evaluates Geddes" section as a "convenient abstract" to describe the characteristics of a region. It contains everything

COMPARISON BETWEEN THE SECTION ALONG THE VALLEY OF P. GEDDES (TAKEN FROM CITIES IN EVOLUTION, 1915) AND THE CROSS-SECTION OF THE VALLEY OF THE SMITHSONS (FROM *THE DOORN MANIFESTO*, 1954, REVISED FOR THE SECOND EDITION OF THE TEAM 10 PRIMER).

needed to live and work together, it can be applied anywhere and represents a system of relations and mutual dependencies between the parts, "organic and self-contained regions" (HILBERSEIMER 1949: 89)[8]. The section contains other potentialities: though not explicitly included among the group of the "thinking machines", it belongs there in its ambitions and due to the type of logical process it develops. Geddes' diagrams are devices that put into circular relation fields that are generally separated. That acts on at least two levels. First, it questions the type of relationship that is established between words and objects arranged inside proximate (two- or three-dimensional) spaces. Next, it defines a new space that contains all the terms or objects present in the diagram. Thus the section can be read: as assertion of the "region" as the new scale of comprehension of the relations between man and his environment; as investigation of the passage from one type of settlement to another, and of their necessity in relation to the characteristics of the territory. Just as the river transports to the sea materials that belong to its previous portions, so every "complex community" is modified by those that came before it. Similarities to the vegetable world intervene, connected with the study of the French biologist Charles Flahault (WELTER 2002: 61), who, at the start of the 1900s, researched associations among plants. The city-country relationship is seen in terms of integration and correct distribution of land use, rather than in opposition; adaptation, evolution and environmental determinism; and cooperation and hierarchy instead of the battle for survival. The section also makes it possible to rethink certain fundamental controversies that crossed the formation of modern urbanism. In January 1954, the section, in this case cutting across the valley, reappears in the report of the meeting at Doorn of the group of young architects participating at the CIAM congresses, who would later constitute *Team 10*. The report titled "Statement on Habitat", published in the second edition of the *Team 10 Primer* (1974) by Alison and Peter Smithson, takes its cue from Geddes' diagram and the idea that the design of the city has to simultaneously focus on community and on environment, comprising their different characteristics. This concept is represented in the diagram "Scale of Association" that returns to and articulates the model of relations as well, with a more radical emphasis on the community as support for living, but rejecting the idea of the "neighborhood unit" as the privileged place of relations (VAN DEN HEUVEL 2000: 43). At Point Three of the manifesto, the authors write: "Habitat is concerned with a particular house in the particular type of community"; each community defines a specific relationship with the environment. The need to produce "convenient" communities in the different points of the section outlines a new program for architecture and urban planning after the Athens Charter. The various forms of clusters placed along the section drive the study of the groupings of the houses.

The term "habitat" — introduced in the modern urban planning debate by Le Corbusier during the course of CIAM 7 in Bergamo (1949), where he urged the preparation of a "habitat charter" — becomes a reference point for many. In 1952, in the preparatory seminar of CIAM 8 at Sigtuna, Candilis presents a diagram for concentric centers with the title *Habitat*, distinguishing between *environnement immédiat* and *environnement urbanistique* around human beings and their lodgings. The concept of habitat sets out to get beyond functional division and particularly the concept of the residential zone, investigating spaces that could be utilized on different scales. A cultural and anthropological issue, the concept of *habitat* developed by Candilis is strongly influenced by French colonial policies in Morocco, where Candilis works, that experiment with the idea of the *Habitat adapté* in relation to different groups and populations (Avermaete 2005: 142).

In the manifesto of Doorn in 1954, we see the return of a form of idealization of Geddes' section that expressed the problematic relations between city and territory and between the different environments; their rigidity, dependence, and strongly ranked status, with the city against the background, absorb what the territory produces. The ideology of the community did not seem to be central in Geddes, who instead represents the relationships of production, but it takes on the character of necessity in the diagram of the Smithsons, modifying the interpretation of the Scottish scholar and reformulating it in a completely altered key. The two diagrams share a territorial hypothesis of urbanity, the idea of a city-region, or of a city in the landscape, that was to become a constant concern of urbanism during the entire 20th century.

REDRAWING OF THE DIAGRAM OF THE "GREEN CITY" NEAR MOSCOW BY GINZBURG AND BARSCH, 1930, AND OF THE DISURBANIST DIAGRAM OF SCHEMATIC DISTRIBUTION OF FACILITIES AND SERVICES (BOTH TAKEN FROM KOPP, A., 1967, *VILLE ET RÉVOLUTION*).

The question of the new form of the city is at the center of many diagrams that pace the first decades of the 1900s. It is posed in terms of its breakdown in the drawings of Bruno Taut (1920) and the disurbanists, who, along the lines of the positions of the Trotskyist sociologist Mikhail Okhitovich, conceive of the city as a mobile entity based on an efficient system of communications. Unlike Soria's diagrams, which are almost exclusively devoted to the city of residence, those of the Soviet "urbanists" and "disurbanists", in opposition regarding the nature of the socialist city, confront the new scale of the modern industrial city in all its complexity. They propose two different principles: organization in bands, in the case of the disurbanists, seen in many diagrams of the same period or somewhat earlier, such as those of René Braem in 1934 (Strauven 1985); and point by point, concentrated and collective settlements along the infrastructures, in the case of the urbanists (Parkins 1953).

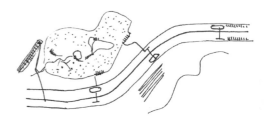

de Town and hinolehm.

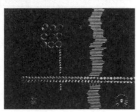

"RHYTHM IS A CREATIVE MEANS OF COMPOSITION". REPRESENTATIONS AND QUOTATION TAKEN FROM VARIOUS AUTHORS, IDEE PER LA CITTÀ COMUNISTA, 1968.

The reflections of the disurbanists, but also those of the urbanists, inserted in the debate on the construction of the "socialist city", came in the wake of the 25th Congress of the Communist Party of the Soviet Union, in December 1927, devoted to the construction of the first Five-Year Plan (1928-1932). Published in 1929, the Plan called for the construction of 200 industrial cities and 1000 rural centers (KHAN-MAGOMEDOV 1987: 284); the reference to Howard still present after the revolution is by now a thing of the past. The critique of the city in Marxist circles can be summed up in certain famous statements of Engels[9], Marx, and Lenin[10] that form the backdrop for the radical choices of the disurbanists: the resolution of the contradictions between city and country would require the elimination, though long and laborious, of urban centers. Considering the model of the capitalist city that is impossible to reform, the disurbanists propose replacing it with ribbon-like settlement structures that follow the road arteries and the trails of energy distribution, spreading across the whole territory. Okhitovich, influenced by the reading of the autobiography of Henry Ford and the possibility of distributed production, defines the city as a human entity determined from a social viewpoint and not rooted in the territory, a process rather than a static set of objects. He calls this *disurbanism* or *disurbanization*. It is ribbon settlements (COOKE 2001), and not linear cities, that are generated by linear infrastructures, with mobile prefabricated homes and shared facilities for socialist life, rhythmically arranged along the axes.

The theoretical system developed by Okhitovich takes on visible form in the diagrammatic project of 1930 for a "Green City" near Moscow, by Ginzburg and Barshch, and in a famous diagram that illustrates the rhythms created by the presence of equipment along the linear ribbons. Some of the ideas contained therein are taken up again in the project for the new settlement of Magnitogorsk by Leonidov in that same period, and in the essay by Miljutin *Sotsgorod: The Problem of Building Socialist Cities* (1930), which proposes a linear city model. Closely observing the two diagrams of the disurbanists and reading the translations of certain passages from Okhitovich, Ginzburg and Barshch, translated by Kopp (1967), the ideas seem richer than has been suggested by many interpretations, starting with that of Kopp himself who complains of the loss of a sense of reality in rigorous architects like Ginzburg and Barshch, or that of Le Corbusier who pokes fun at these works in the *La Ville Radieuse*. In effect, the ribbon cities capture the moment of a fundamental transformation of the Russian territory that lent itself to infrastructure on an

REINTERPRETATION OF THE "GREEN CITY" PROJECT FOR MOSCOW, 1930. STARTING CONDITION, DECENTRALIZATION OF PRODUCTIVE, SCIENTIFIC AND ADMINISTRATIVE ACTIVITIES, DECENTRALIZATION OF RESIDENTIAL AREAS, CONCENTRATION ALONG LARGE RADIAL CIRCULATION AXES, MAINTAINING OF GENERATED VOIDS; THE CITY BECOMES A PARK COMPOSED OF MONUMENTS, CULTURAL AND ADMINISTRATIVE FACILITIES.

  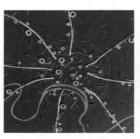 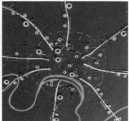

unprecedented scale and was electrified thanks to the construction of distributed hydroelectric plants. While the urbanists propose collective lodgings and episodes of density organized like a rosary, the disurbanists, in their diagrams, represent a condition of even more extensive infrastructuring, which nevertheless is not utterly unrealistic. A grid would cover the entire territory, as Okhitovich asserted (KOPP 1970: 173). The idea of dispersed settlement that had been initially judged as coherent with communist positions, as a vehicle of emancipation of the Russian peasant, is soon pushed aside and deemed unfeasible due to its high infrastructural costs.

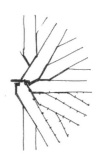

A variation and reworking of the concept of the linear or ribbon city can be seen in the diagrams proposed by Hilberseimer, among others, before and after his move to the United States. In *The New City* (1944), Hilberseimer develops certain positions on the new scale and form of the city, confirmed in later books in which he takes up the same ideas and uses the same references. His first conviction is that the tendency towards decentralization is in progress and is inevitable. The diagrams structuring the dispersion also represent a new relationship between the individual and the society. We find echoes of disurbanist thinking and of the words of Henry Ford, who in the 1920s pointed to a lack of necessity for industrial concentration (FORD 1922), a perspective later taken up by Wright and, to some extent, by Franklin D. Roosevelt in the New Deal of the 1930s.

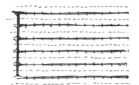

The second conviction has to do with the use of the linear or "ribbon" structure, since it lends itself to experimentation with a new hybrid condition of integration of industry and agriculture. The radiocentric city is totally rejected. The third point addresses the new relationships between industrial and agricultural production; the references are to Ford, but also to Kropotkin (HILBERSEIMER 1949: 82; KROPOTKIN 1899). Among the diagrams of the linear city, appears Wright's Broadacre City, a low-density city organized along a highway, along the same lines as the diagram proposed by Hilberseimer himself (HILBERSEIMER 1944: 71). To support the choice of the linear city, as opposed to the radiocentric and radioconcentric settlement, Hilberseimer uses the diagrams of Hans Ludwig Sierks[11] and Peter Friedrich[12] that compare the efficiency of the two models from the viewpoint of rail infrastructure (HILBERSEIMER 1944: 73), reaching the conclusion that the ribbon development is more efficient than the concentric model.

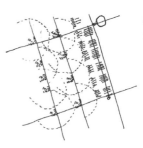

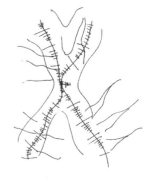

Finally, the design of the city, according to Hilberseimer, is a problem of performance and requires correct formulation in terms of orientation, wind direction, and the risk of fires. The city, as in the visions of the disurbanists, becomes part of the landscape: "the city will be within the landscape and the landscape within the city" (HILBERSEIMER 1944: 126)[13].

In the subsequent text, *The New Regional Pattern*, Hilberseimer uses the term "landscape" in much more sweeping ways, as a synonym for "environment", pro-

TOP TO BOTTOM, REDRAWING OF THE DIAGRAMS: H.L. SIERKS "CENTRALIZED TRAFFIC SYSTEM" (HILBERSEIMER 1944); P. FRIEDRICH'S "TRAFFIC SYSTEM IN RIBBON DEVELOPMENT" (HILBERSEIMER 1944); "PLANNING SYSTEM" (HILBERSEIMER 1944); "RURAL PLANNING SYSTEM" APPLIED TO AN IRRIGATED AREA SURROUNDED BY PASTURES (HILBERSEIMER 1949).

posing a "comprehensive" approach that begins with reflection on the ecological, integrated and systemic cycle as the basis of territorial design. Of the three urban planning systems represented in the diagrams, the first is "predominantly urban", the second is "predominantly rural", and the third is a combination of the first two, conceived for highly industrialized and hybrid situations. While the first diagram reprises those published in *The New City*, the rural diagram introduces the idea of integration between agricultural and industrial production, then developed in the successive drawings.

The diagram is presented in its abstract version, without reference to a precise place or to physical characteristics – rivers or hills – but it is also inserted in a context with the clear, explicit aim of demonstrating the flexibility of the idea, its capacity to adapt to the forms of the territory and the different conditions, just as in the diagrams of the disurbanists. For Hilberseimer, the diagram is the way to convey "abstractions", while ideas and theories are the starting point "for discovery of our methods of work" (HILBERSEIMER 1944: 128). The diagram is the way of establishing a dialogue with the outside world; the discussion around the design of cities and territories that it launches is not confined to a disciplinary debate and cannot be said to be concluded without a much wider debate.

The materials of the project: concepts of the void

The construction of an urban territory becomes the opportunity to approach the theme of social, economic, and cultural imbalances between city and country, or between territories, and brings the geographical scale – the new functional and symbolic role of open space – into the foreground. Penetrators, wedges, green belts and corridors arm the modern urbanist with new analogies and metaphors that introduce unprecedented design figures that from this point on, for over a century, structure his/her projects. In the sketches and diagrams of Le Corbusier, open space never manages to detach itself from the background, to take on its own figure. In *Les trois établissements humains*, his most complete effort of imagination of new territorial concepts, though less suggestive than the project for Algiers or the *Ville Radieuse*, three separate settlement and structural types are represented that determine different relationships between full and empty zones, but rest on a formless surface. Only later, in the project for Chandigarh, based on an earlier plan developed by Albert Mayer (LONERO 2005), does the open space take on forms based on criteria of continuity and connected to the various practices of the movement. By the 1950s, these ideas were part of a shared legacy.

However, the, indifference to the form of the territory and to the structural and organizing capacity of open space is not shared by all the modern architects and urbanists. In Germany, in particular, a focus on the ecological functioning of the territory can be found in the work of landscape architects like Leberecht Migge, who takes part in the preparation of the most interesting projects done in Berlin and Frankfurt in the second half of the 1930s. The profoundly anti-urban picturesque tradition and

THE LINEAR CITY: A PORTION OF
THE METROPOLITAN REGION OF
BARCELONA.

the deeply rooted sense of nature provide the ideological and conceptual underpin-
nings for an emphasis on open space and its lasting charm. Certain concepts of the
void, like that of the green belt, are integrated to the point of merging with other concepts, for example those of the cellular tissue and the archipelago, producing some
of the most powerful and lasting diagrams of modern urban design.

Grüne Gürtel.   In 1874 in Leipzig, the book by Arminius on the metropolis focuses
on the question of housing and reflects on the need for green areas[14] in the form
of a belt of woods and meadows accessible to all citizens within the radius of one
mile, with a depth of one mile (1600 m). A place of contact with nature, recreation
and cultivation of vegetables, the green belt – the Countess Dohna-Poninski, *alias*
Arminius, points out – does not necessarily have a circular form, and should allow
for only one fifth of the land to be developed. Rudolf Eberstadt publishes a diagram in
his manual of 1909. Hegemann discusses it in *Das steinerne Berlin* (1930), indicating an Elizabethan law of 1580 against the practice of subletting that imposed the
construction of new housing units at a distance of about five kilometers from the

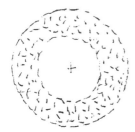

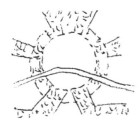

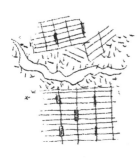

GREEN BELT: REDESIGN OF THE
CONCEPT OF ARMINIUS, 1874.

GREEN BELT: REDESIGN OF THE
GREEN SPACES IN THE DIAGRAM
OF ARMINIUS REFERENCED BY
EBE-RSTAD, 1909.

ADELAIDE: SCHEMA OF THE CITY
FOUNDED IN 1836.

existing city as a precursor of the idea of the *grüne Gürtel*. Hegemann himself presumes that the countess, acquainted with Lord Shaftesbury, participant in the housing reform movement in England, was not aware of the Elizabethan legislation, though she must have known about the plan of the city of Adelaide from 1836, surrounded by a band of "Park Lands" 800 meters deep. In the design of Adelaide, also a model for the Garden City of Howard, a green belt is freed of the need for a circular plan of the city organized along radial spokes, as happens instead in successive concepts, outside the hierarchical hypothesis of relations between the big city and the suburbs beyond the belt, and without forming a ring. The green belt combined with a grid layout and the limited size of foundation cities gives rise, instead, in colonial cities, to a Benthamian panoptikon that organizes and concentrates the population in separate groups – in places visible from all sides and from the outside. The social engineering experiment theorized by Wakefield[15] with the colonization of Australia and New Zealand, which responded to problems of overpopulation and congestion in large cities, takes form with physical elements designed to create equal opportunities, represented by the grid and the "common lands" or "parklands" around it.

From colonizers like Sir Robert Montgomery with his diagram from 1717 of green belts around all the districts of the Margravate of Azilia in Georgia (never built), to social reformers engaged in the struggle for the abolition of slavery like Granville Sharp, who imagines founding a colony of freed slaves in Africa[16], all the way to T.J. Maslen, the pseudonym of the retired naval officer Allen Gardiner, who in 1828 in *The Friend of Australia* writes of entering the city through "a belt of park", as well as the New Lanark of Robert Owen... It is evident that ideals of equity, alongside those of enclosure and separation, are enacted in the new city through the concept of the green belt.

Cellular diagrams

All the social and political ambiguities connected with the green belt diagram find their way into the cellular diagrams that set an idea of the city, often a large one, on a spongy, dense void, a space of absence, of distance and separation. It is also a place of relations, a territory of conquest, and space of freedom. It contrasts the severe order of the individual cells and is fundamental to their functioning.

In the passage from the 1800s to the 1900s in Germany, the construction of the metropolis becomes an internal colonization of the territory in pursuit of a new balance, as opposed to external colonization. The emigration to America, Hegemann recalls, was indicated by Goethe in *Wilhelm Meister's Journeyman Years* as the only way to get out of the disastrous urban housing situation. In response to Goethe's suggestions and after having studied the condition of the working class in England, Victor Aimé Huber advances, in 1848, the theory of "internal occupation" (*innere Ansiedlung*): "in suitable points, in the suburbs of large cities, at the intersections of major railway lines, in a suitable position between rural properties, villages, factories [...]"[17]. The manual of Eberstadt (*Handbuch des Wohnungswesens und der Wohnungsfrage*, 1909) also contains a chapter on this theme. The diagrams of Erich Gloeden are perhaps the most extreme representation of the cellular project. Hilberseimer carefully studies the diagrams contained in *Inflation der Gross-Städte* (Gloeden 1923) and publishes them in *The New City* (1944), comparing them to those of Unwin for Greater London and pointing out the differences. More recently Paolo Sica, in his monumental *Storia dell'urbanistica, II Novecento*, has gone back to those diagrams, defining them as schemes of colonization, and that of Gloeden as "a singular proposal" (Sica 1978: 170). In both cases, the references to Gloeden are short and the legends of the diagrams are not included.

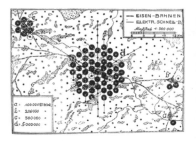

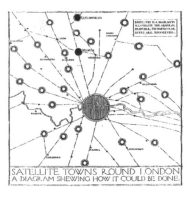

According to Hilberseimer, Unwin's satellite cities around a main center, inspired by Howard (about 6,000 inhabitants for each of the three residential sectors, 1/4 of the area set aside for industry, and again, in this case, integration between industry and agriculture), fail to clearly resolve the issue of the growth of the city. Gloeden's diagrams, on the other hand, suppose the existence of a main city inside the metropolis; the growth is by gemmation around the initial nucleus, and at a certain distance. The whole is more dense and larger than in Unwin's proposals. Each cell, with its own specialization of no more than 100,000 inhabitants and a radius of no more than three fourths of a mile (1.2 kms, still a walkable distance) has its own physiognomy. A railway connects each settlement, and the process of gemmation is potentially infinite. The only critical note has to do with the position of the industrial facilities at the center of the settlement, leading to grave consequences for air quality, a problem to which the famous "fan diagrams" in *The New City* attempt to provide a response. The result, according to Hilberseimer, is "a new city type" (Hilberseimer 1944: 63).

E. GLOEDEN, SCHEMA OF THE PLAN OF A SETTLEMENT, TAKEN FROM *INFLATION DER GROSSTÄDTE*, 1923.

R. UNWIN, B. PARKER, "DIAGRAM OF THE SATELLITE CITIES AROUND LONDON", 1903 (HILBERSEIMER 1944).

If we examine Gloeden's diagrams, the observations of Hilberseimer can be remarkably enhanced. First of all we rediscover a text, *Die Inflation der Gross-Städte*, published in 1923 by Erich Gloeden[18], a young Jewish architect who belonged to the Loevy family of Berlin, owners of the foundry of the same name since 1855. Thanks to Behrens, the Loevys had produced the bronze inscription placed on the Reichstag in Berlin during World War I ("Dem Deutschen Volke", "in the name of the German people"), and they were protected by architects of the caliber of Gropius and Mies van der Rohe. Erich Gloeden was born in 1888, earned a degree in architecture in Dresden in 1915 (he probably studied with Friedrich Ostendorff), and in 1918 he was baptized and changed his name. He was sentenced to death

CONTENTS

CHAPTER 1: THE STATE OF AFFAIRS
The problem of traffic in big cities – The natural organization of rural settlements – The unnatural contemporary expansion of cities – The impossibility of an orderly project – The search for a remedy through transport systems – The unrentable city – The failure of organization of the project – Remedies for housing demand – The Space-Time concept

CHAPTER 2: THE OBJECTIVE
I Decentralization in keeping with an agricultural model – Maximum radius of action – The concept of inflation in the construction of settlements – Ideal economic framework of the space-time mega-cluster – Systematic separation of the economic branch in keeping with the medieval model – Total elimination of automotive traffic inside the city – Elimination of traffic by light metropolitan rail systems – Cellular organization of smaller urban clusters – Urban expansion as differentiation of labor.
II Settlements for jurists and retired people (two examples) – Adult youth – The machine and crafts – The income of the city – Channeling – Infrastructural connections with neighboring cells.
III Parallels with the natural sciences – Inflation of large cities in antiquity and during the Middle Ages – The three phases of expansion of the city in the 19th century – The fourth phase.
IV Settlement forms and residential density – Iso-lated productive units – The city of one hundred thousand – Expectations regarding today's technicians – Statistical comparison of settlement densities – The apparent paradox and its solution – The embryo of the city – The optimal number of stories and the economic optimum – The center of commerce and the tower building – Residential ring and two-story buildings – The debate on vertical or planar construction – Calculation of the solution to this problem according to the Space-Time concept.
V Transformation of the external character of the city – Ideal artistic representation of the space-time mega-cluster – Constant radius as replacement of medieval ring fortifications – Transformation of the internal characteristics of the city – Visible economic responsibility – The evident economics of level of appreciation.
VI The project according to a four-dimensional concept.

CHAPTER 3: THE PATH
I Medieval urban economics in the 19th century and the consequences – Aspiration to Taylorism – Increase of salaries in the past – Stabilization – Different urban concentrations – Invisible capitals.
II Development of ancient cultures – The ancient city at the crossroads – Development of new settlements – Expropriation of the entire territory of the village – Economic privatization of municipal administration – The four constructed areas – The new style of the time.

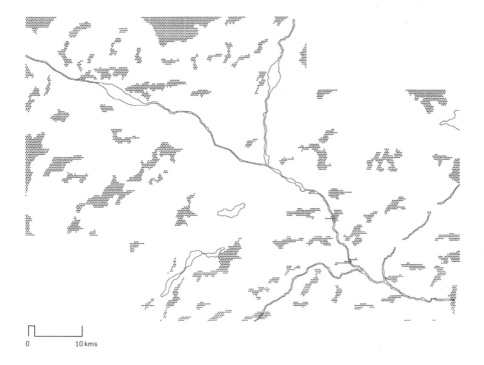

0   10 kms

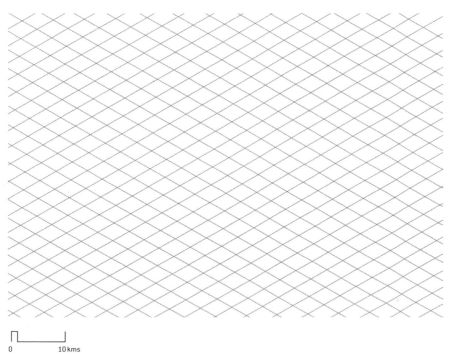

0   10 kms

DECONSTRUCTION OF THE
DIAGRAM "SCHEMA OF THE PLAN
OF A SETTLEMENT", PLATE 1: THE
"REASONS" OF THE TERRITORY
(WATER AND FORESTS); THE
"REASONS" OF GEOMETRY
(GLOEDEN 1923).

together with his wife and mother-in-law by the People's Court in 1944 for having concealed, by request of Hans Ludwig Sierks, General Lindemann, who was involved in the attempted assassination of Hitler[19].

The caption of the first of the series of plates ("Scheme of the plan of a settlement") reads: "The chart contains 150 rural settlements and 66 urban settlements", organized in the form of 4 clusters. A dense railway and streetcar network (*elektr.schnell-bahnen*) equipped with large interchange platforms structures the territory[20]. Each cluster, at the side or the center, has one of the large railway areas (always of the same size): the smallest cluster has a single settlement of 100,000 inhabitants joined by rural settlements, while the largest has about 5 million inhabitants (in the 1920s, Berlin had a population of roughly four million), and is served by two platforms. While the railway serves large industry and the urban centers, the territorial streetcar network covers differentiated routes that permit the isotropic organization of the centers.

Despite their apparent simplicity, Gloeden's diagrams propose solutions of some interest that emerge as soon as one attempts to redraw them. The first diagram, on a territorial scale, is constructed on two parallel planes, following two different rationales that express the "reasons of the territory" and the "reasons of geometry". A territory, though abstract, does exist; its form can be grasped starting with the presence of water, since the topographical information is not provided. There is an underlying geometry, a rhomboid pattern that forms the basis of the design of the cellular settlement. The form of the territory guides its layout; the design of the railway networks and the location of the large industrial platforms are generally placed where water and rail meet. Geometry, on the other hand, guides the isotropic organization of the cells, defining the trajectories of the tramways that connect them and contribute to articulating the isotropic distribution of the centers. Several old villages remain as exceptions nestled in the new fabric. The two "reasons", or rationales, proceed in parallel and do not claim to shape the whole. When they overlap, a random, fortuitous encounter is produced, as in the definition of beauty of Isidore Ducasse, better known as Comte de Lautréamont (1869): different situations, in relation to the landscape or the degree of accessibility.

CHAPTER 2: THE PURPOSE

[...]. The problem afflicting the contemporary metropolis is the difficulty and duration of movements from the place of residence to the place of work. The remedy is a return to decentralization in keeping with a rural/agricultural model of proximity between dwelling and workplace. This would allow the population to live close to places of work, eliminating the problem of the overloading of mass transit.

[...]. The radius of action of 15 minutes (based on the rural village) ought to become a fundamental point in the design of big cities. Where it cannot be complied with, urban cells should be provided that are independent from the viewpoint of organization and management, though they are seen as part of the whole with the rest of the urban settlement. Each inhabitant should be able to reach his place of work or school in a maximum of 10-15 minutes on foot, to limit the use of public transportation.

Any urban settlement that does not comply with these parameters is suffering from the problem of inflation.

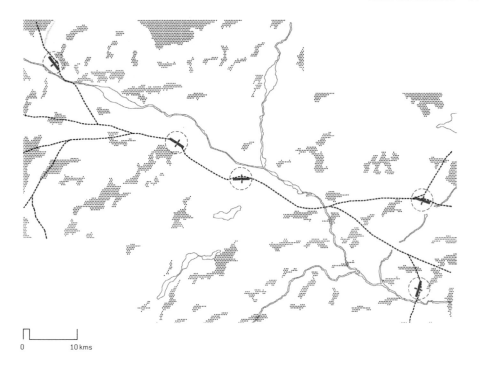

0   10 kms

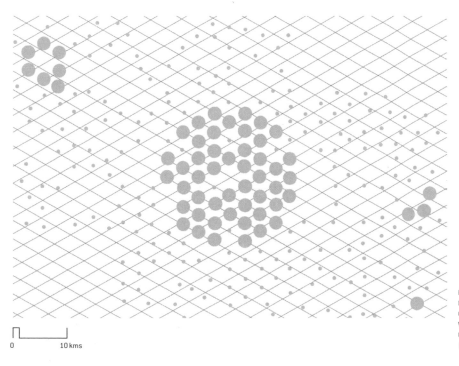

0   10 kms

DECONSTRUCTION OF THE
DIAGRAM "SCHEMA OF THE PLAN
OF A SETTLEMENT", PLATE 1:
WATER AND IRON; GEOMETRY,
URBAN AND RURAL CELLS
(GLOEDEN 1923).

The void, or the space between the cells, is what defines the scale of the big city in which — and in an utterly surprising way — no hierarchically organized street network exists. The cells are not "sectors" separated from each other by streets of busier traffic. Only the railway and the streetcar lines cross the isotropic organization, inserting certain preferred directions. What might seem like a drawback is actually a fundamental absence. The big city is conceived to enable residents to do without cars[21]; the lack of sectorialization and of a ranking of the parts breaks up the traditional center-outskirts relationship and opens to an interpretation of the diagram à la Alexander (1965): the city of many millions of inhabitants is not a tree, but a voluntary cooperation of islands that share the public transport infrastructure and the areas of production. All this is very distant from other interpretations of the cellular concept, where it is always associated with a ranked structure. One example are the criticisms advanced by Alexander regarding the diagram that describes the "villages" of which London is composed in the *County of London Plan* (1943) of Abercrombie and Forshaw (who intended to accentuate segregation in order to reinforce identity).

The density proposed by Gloeden is higher than that of the London region (250 inhabitants per hectare) and lower that that of Berlin at the time, a constant point of reference for the author: "The population density of Berlin is 33,300 inhabitants per square kilometer, which compared to our hypothesis would mean 150,000 inhabitants in one cell. But we should take into account that the city of Berlin is overloaded and that most of its residential buildings are overcrowded, without air and light. Optimal density is achieved by eliminating everything that is in excess inside the cell, conserving residence and work inside it, while hospitals, monasteries, schools, and prisons are positioned in the green belt or in specialized cells"[22].

The production areas are incorporated inside the railway platforms, of immense size, as much as 5 kilometers in length, in which oblong, spindle-shaped urban centers are placed, equipped to serve production and also positioned along the tracks. The image evokes the description of the *Stahlstadt* of Jules Verne in the novel *Les 500 millions de la Bégum* published after the Franco-Prussian War (VERNE 1879): a landscape of steel, the "city of the railroad", at the service of production and the movement of goods, but also of the rationalization of the location of production facilities, taking advantage of waterways connected to the railroad through ports and river piers.

REDESIGN OF THE PLAN FOR GREATER HELSINKI BY ELIEL SAARINEN AND BERTEL JUNG, 1918.

The void between one cell and the next reminds us of the schemes of Eliel Saarinen and Bertel Jung in 1918 for Greater Helsinki, but clarifies with much greater precision the role of the green belt proposed by Arminius in the construction of the metropolis, cited in the first chapter of *Die Inflation der Gross-Städte*. Gloeden uses the term *grüne Gürtel* as a device that organizes the big city for many millions of inhabitants. The open space is its excipient. The distance

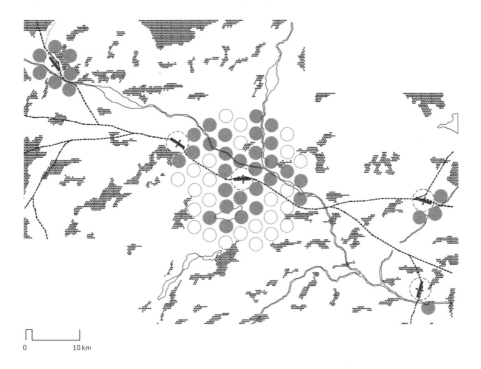

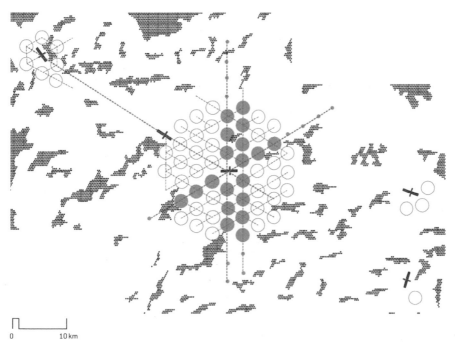

DECONSTRUCTION OF THE
DIAGRAM "SCHEMA OF THE PLAN
OF A SETTLEMENT", PLATE 1:
ISOTROPY *VERSUS* ANISOTROPY:
CELLS NEAR THE RIVER OR
THE RAIL LINE; PATTERN OF
STREETCARS AND CELLS NEAR
TERRITORIAL STREETCAR LINES
(GLOEDEN 1923).

CHAPTER 2: THE PURPOSE

I

To make this expectation become reality, I propose an ideal model of the city, as follows: at the center are the cities of labor and administration, and buildings with light metropolitan rail stations are at the basement level. Around them the residential buildings are grouped, all the way to the outskirts, which can be conveniently reached in 10-15 minutes. The residences of the highest quality are along the external margin, far from the traffic, with their fronts oriented towards the green belt composed of fields and woods.

The green belt separates the cells and has a minimum width of 500 meters. Sports fields and playgrounds periodically repeat between the cells. The large block in the green belt contains elementary and middle schools, easily reached in 15 minutes from the three neighboring cells. Thus the center of education sits diametrically opposite the center of traffic, distant from it and in an excellent position.

What will the center of the city – the core – contain? All these centers contain theaters, spas, the post office, a department store, shops, restaurants, power utilities, the local administration, and, finally, the station. Until today production districts have been located chaotically inside cities. In the metropolis of the future, this will no longer happen: textile cells, bank cells, cells for commerce, exports, administration. At the edges of the cells there are always houses for the workers. In the capitals there will be even more specialized cells, with ministries, parliaments, and museums. If a worker has to move to a company in another cell, while keeping his residence the same, he just has to walk 35 minutes across the belt of fields to reach his new place of work.

Most people move out of their cell only for exceptional reasons. The principle always remains the same: few move towards many, not vice versa. Those who have the necessity of moving from one cell to another can do so thanks to convenient, rapid means of transportation.

Each cell has 9 kms of track. Given that the use of these transport systems is occasional, they have the advantage of a constant number of users, both day and night. Since no transport system is needed inside the cell, a couple of horse-drawn carriages will suffice, for the transport of the sick. Hotels and restaurants are placed in the direct vicinity of the light metropolitan rail line at the center of the cell, and some freight cars permit easy transport between different cells.

All snags between space, time, and energy are reduced to a minimum.

The central station and the secondary stations contain the management, depots, ports, and lodgings for the railway workers. The light metropolitan rail lines connect each cell, passing over and under the platforms of the station to connect the station directly to the center of the cells.

The city is therefore quieter, less chaotic, and, above all, it permits optimization of the time of its inhabitants, something that has become very difficult in the contemporary city. (cont.)

II

At the center of the settlement complex, there is an old village of great environmental value. The village is conserved as a model agricultural business.

Around the village, on the outer margin, a strip of urban vegetable gardens is developed for the inhabitants of the neighboring cells. The size of the drainage canals can be reduced, since the strip of urban gardens can be used as a basin for gathering and purification of the wastewater from the nearby cells. (…)

III

This procedure simply imitates what already happens in nature. Amoebae also separate after a given period of life to give rise to a new living being, rather than growing infinitely.

The large cities of antiquity were not able to solve the problem of cellular separation. They were not capable of expanding, because they did not have the suitable means of transport. (…)

Also in the medieval city, the problem of decentralization was not perceived. Fortifications determined stable borders.

The three phases of the development of the 19th-century city were: 1. concentric expansion around the historical center; 2. in place of gates, railroad stations were built, and growth became radial; 3. neighborhoods grew up, disconnected from each other, along the regional rail lines.

This third phase needs to become much more organic.

Also in the fourth phase, the new districts will develop along the stops of the railroad, but unlike what happens today they will not be like drops around a center – no longer satellites. The old center of the city will retreat like a primus inter pares.

The city will therefore be formed by many centers, one beside the other.

IV

In what way and with what density can the constant surface of each cell be constructed?

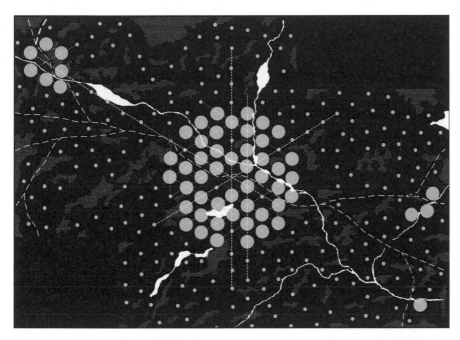

INTERPRETATIVE REDESIGN OF
THE DIAGRAM "SCHEMA OF THE
PLAN OF A SETTLEMENT",
PLATE 1 (GLOEDEN 1923).

What is the density per hectare and what are the optimal building types?

From the reasoning expressed thus far, the optimal radius of action appears to be 1250 m, a range that in European rural conurbations has been a constant dimension for centuries. (...)

The inhabitants should have the same area at their disposition as in country villages.

The residential density will be calculated to have 100,000 inhabitants, a number that is well suited for a radius of 1250 m. This makes it possible to obtain a residential density of 220 inhabitants per hectare, with 45 m2 for each inhabitant. The total surface of a cell with radius 1250 m will be 491 hectares.

Optimal density is also obtained through the elimination of all surplus found inside the cell. Only residences and workplaces are located there, while hospitals, monasteries, schools, and prisons are located in the green belt or in specialized cells.

This gains lots of air and open space. Of course the cell does not contain parks, but it should be kept in mind that in 15 minutes, it is possible to reach the open spaces of the green belt.

A park would have no reason to exist inside the cell, while in contemporary cities parks are fundamental. (...)

The green belts constitute 100% of the open space required by the city. (...)

V

(...) From the viewpoint of mobility, it is interesting to note that when crossing the city, one does not have to pass between buildings, as happens today, but through woods and meadows. The silhouettes of the buildings can be seen in the distance. A monumental zone here, a factory there, residences further on. The green belt grants unity, like the walls of the medieval city or the bastions of the Renaissance.

The green continues into the center of the city, and from the center of the city, it permeates the courtyards of the buildings.

It is an elastic belt that gathers all the diversities of the city, separates dwellings and places of work, provides large open areas near the city center, and at the same time contains the geometric connections between the external radial settlements and the internal checkerboard systems. (...)

At the center of the cells, there are monumental squares and gathering places for social life – not infrastructural nodes. (...)

Until the 19th century, all cities had an organic connection to the landscape. Even when transformed into promenades, boundary walls and bastions constituted a visual border that framed the landscape. They gave the design of the city a direction and an end, orienting the settlement towards the landscape. (...)

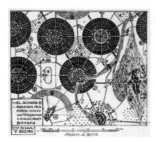

ELEMENTS OF THE CITY

A. Administrative city. Includes all the administrative offices of the state, province, and districts of all the circular settlement areas. It also includes parliaments, embassies, etc.

B. City of culture. Academies, universities, polytechnic institutes, museums, state libraries.

C. City of commerce and large retailing. Only offices at the center.

D. Medieval city reconstructed along the river using the local artisanal culture. Professional schools, crafts workshops, workshops in which to teach trades to young people, shops for retailing quality products.

E. City of industry. The green belt between the city of work and the city of residence contains administrative and cultural facilities. At the center: industrial buildings of ten stories for industry and handling goods.

F. City of the railroad

G. Central slaughterhouse and stalls, for the entire settlement area. At the center there are buildings for the employees of the slaughterhouse and public buildings constructed on the ruins of an old village.

H. Large open area around an old settlement now incorporated in the urban system. There are small urban vegetable gardens at the edge of the city. The old settlement is thus transformed into a model to conserve for the city.

ELEMENTS OF THE GREEN BELTS

1. Complex of buildings for elementary and middle schools
2. Hospitals
3. Cemeteries
4. Prisons
5. Track and field sports facilities
6. Cavalry barracks
7. Playground and exercise area
8. Urban vegetable gardens

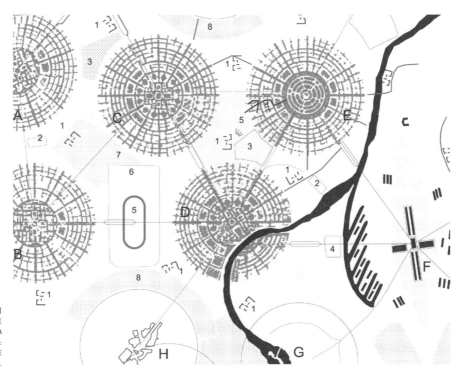

(NEXT PAGE)
REINTERPRETATION OF THE
"SCHEMA OF THE PLAN OF A
SETTLEMENT", PLATES 1 AND 2:
LA CINTURA VERDE
(GLOEDEN 1923).

between center and center is about 800 meters, crossed by tree-lined avenues that connect them. The avenues link back to the tradition of the *promenades*, long routes in the continuous park of the big city, described by Stübben in his *Der Städtebau* (1890), a text that appears amongst Gloeden's references. The latter writes: "From the viewpoint of mobility, it is interesting to notice how, crossing the city, one need to pass between buildings, as happens today; one crosses woods and meadows, while in the distance the silhouettes of the buildings can be seen. Here a monumental district, there a factory, with residences further on. The green belt grants unity, like the walls of the medieval city, or the fortifications of the Renaissance"[23]. There are many versions of this space, a true equipped platform of the metropolis – from sports centers to urban vegetable gardens that protect the villages incorporated in the new cellular tissue, to woods and cemeteries distanced from the more densely inhabited areas. The great void is continuous, always nearby, making even the forests, rivers, and swamps domestic, crossed by paths, incisions inside the nuclei that adapt to their presence. The apparently homogeneous and isotropic space conceals continuous variations. Each center has slightly different characteristics from the others, with its own formal recognizability and functional specialization.

The separation of the form of the infrastructure from that of the city distinguishes Gloeden's diagrams from those of the linear city; the central role of the railroad and waterway infrastructure is clearly stated, but the city grows according to a rationale that does not harken back to the traditional radioconcentric organization. Though applied in the cells of 100,000 inhabitants, it is rejected in the new urban scale. The void is what defines the dimension of a city inscribed in a diameter of 25 kilometers, but which could be much larger. There is no reason to organize the cells in a peripheral way around a main center that does not exist. Instead, they are linked by a relationship of proximity and complementarity, optimized in the figure of the fabric. The reference is to the amoebae that separate after a certain time to give rise to a new individual. What is represented in the diagram is the solidarity of the group: the almost spontaneous association of different identities is reinforced in the new organism, contrasting hierarchy, and the law of the strongest: "[...] the old center of the city will retreat like a *primus inter pares*"[24]. The diagrams of the central place theory of Christaller come 10 years later (1933)[25].

Nuclearity and the archipelago city.   Reflection on nuclearity marks the whole 20th century, influencing the thinking of architects and urbanists on the growth modes of the city and the role of open space in its configuration. The French geographer Vidal de la Blache also insists, at the start of the century, on colonization as a mode of growth, recalling the periods in which Europe was being occupied from within and referring to this phenomenon as a part of its history. Giovanni Astengo, a number of decades later, indicates some of the cases in the margins of his copy of *Principes de Géographie humaine* (VIDAL DE LA BLACHE 1922). Gemmation by nuclei is opposed to sprawl and combines with growth by attraction and accumulation, like coral. While the influence of the biological sciences is strong, the organicism of Vidal is more metaphorical than substantial. Astengo's notes on the pages of *Principes de Géographie humaine* evoke the theses of Pierre Émile Levasseur on population distribution in relation to points or lines of attraction. Astengo writes in his notes, "fundamental: the nuclearity of spatial distribution of the population. Two possibilities of grouping: lines and nuclei", adding a red asterisk for the phrase "Les hommes ne se sont pas répandus à la façon d'une tache d'huile, ils se sont primitivement assemblés à la façon des coraux" (VIDAL DE LA BLACHE 1922:10). In successive layers different populations – or human floods, as Vidal writes – have accumulated in chosen places, as with coral reefs. The case of Niger, where the villages multiply by gemmation, founding new colonies once the periphery has spread beyond a range of 1,800 meters, piques Astengo's interest. At the bottom of the page, he writes: "urbanization avoided with continuous colonization./an example on which to meditate!/Niger in the avant-garde?"[26].

The viewpoint of the geographer and that of the urbanist, belonging to two separate eras, notice different aspects. While for Vidal it is still the relationship with the ground that determines both the settlement choices and the relationship between productive land and the community it can nourish, for Astengo it is the mode of expansion, interpreted in abstract ways, that appears as common ground between the problems of growth in Italy and the villages of Niger. Actually, what seems to interest Vidal most are not so much the points of density as the intermediate gaps, the voids that require explanations and force a molecular arrangement, defined as a lack of relations. So he calls on the interpretation of Richthofen, from his travel diaries published in 1907, of China as being made of "chambres" and "chambrettes", or of India, citing Henry Sumner Maine, as an "assemblage de fragments". He also evokes the Mediterranean with its large settlement gaps, alongside peaks of density, as in Murgia. At the time of the Industrial Revolution, the distribution of the population "s'offrait déjà comme un palimpseste sur lequel dix siècles d'histoire avaient inscrit bien de rature" (VIDAL DE LA BLACHE 1922:40), to which the great urban expansion adds a new layer. If the Chinese and Indian villages are described as a multitude of little groups or living cells, their pattern of relations ties them to a larger culture, despite their isolation. Starting with the geological metaphor, Vidal reflects on the results of prolonged sedimentation produced by "population floods". Even the European colonization of the globe is thus inserted and legitimized inside the long process of formation of identity. It introduces cohesion where only "scattered materials" previously existed (VIDAL DE LA BLACHE 1922:65).

In modern urban planning, as we know, there are many diagrams on new settlement units and their processes of gemmation and growth. The development of the concept of the "neighborhood unit" belongs to this favorable moment for the cellular analogy that permeates another major disciplinary field that is in a state of formation, urban sociology, together with history, economics, and human geography. In 1925, *The City* is published, a text in which Park, Burgess, and McKenzie include the famous diagram of concentric strips, as opposed to diagrams by sectors (HOYT 1939) and by nuclei (HARRIS and ULLMAN 1945). In all the cases, the models have emerged from investigations conducted on many cities: in the case of Hoyt, for example, 142 cities in the United States are analyzed from the viewpoint of land use and its evolution in time, drawing on the regular patterns that emerge to generate a possible mathematical expression. Diagrams, initially used to describe situations, are overturned into normative models in which a character of necessity is represented[27].

An almost embarrassing application of the diagrams of Gloeden, though with certain important differences, is found in the text *The Ideal Communist City*, published in Moscow in 1966[28] by a group of professors and students from the architecture school. Marxist reflection on "relations in life" and the failure to create a truly "Communist" space form the starting point for the proposal of a new urban environment: the "socio-spatial unit of a new society"[29]. The size of 100,000 inhabitants determines the "New Unit of Settlement" (NUS) conceived for pedestrian movement (20 minutes on foot, as in the case of Gloeden). The linear scheme is criticized for the excessive distance it imposes with respect to the center (the same density and a thickness of 1.5 kms) and tolerated only when a schema of rings is impossible for topographical reasons[30].

THE "ENTIRELY COMPLETED NUCLEI" AND THE "UNIFIED SPACE OF THE NER" FROM VARIOUS AUTHORS, IDEE PER LA CITTÀ COMUNISTA, 1968.

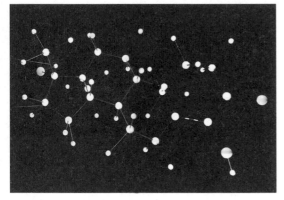

The authors' statements are rigid and final: "The chaotic growth of cities will be replaced by a dynamic system of human settlement composed of fully complete nuclei". The part is finished, concluded, impossible to modify, and, according to the authors, it represents a profound innovation in urban planning culture: "The goal is to transform the whole planet into a unified sociological environment". As in many other visions of those years (in the drawings of Constant, Doxiadis, Friedman, among others), the entire territory becomes the frame of reference, and the settlement is conceived for many millions of inhabitants, in areas rationally planned in terms of tens of kilometers. "NUS is the fundamental unit of that organism. It is the "quantum" of the urban environment, the finite unit, limited by itself and directing itself"[31]. The overall diagram, however, differs from that of Gloeden and counters the large industrial complexes with "free nature", positioning agricultural areas and settlements presumably served by a line of main infrastructures between the two. The distance from nucleus to nucleus equals between one to two diameters of the nucleus, but the role of the empty spaces is the same. The unitary space of the NUS is "a great hall under the open sky".

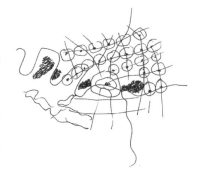

The analogy is with natural and social forms that develop a new organism, without indefinite growth. In this sense, the cell is self-regulating and can be implemented only within socialist conditions[32]. Actually, the principles of the diagram are similar to those of Gloeden: "equal freedom of movement for all", "regulation of distance based on pedestrian movement", "elimination of the dangers of the circulation of motorcars", "green belt". The description of the nucleus eavesdrops widely on Gloeden's *Inflation der Gross-Städte*. The following passage seems almost to mock references not explicitly cited: "The idea of limiting the growth of the community is nothing new in itself. The literature on

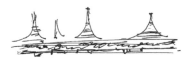

urban planning in capitalist societies is full of good intentions for ending the monstrous expansion of metropolitan areas. Given their social context, however, these ideas, no matter how reasonable, are impossible to put into practice". Finally, the diagram of the NUS is constructed like a sector of Le Corbusier.

REDESIGN OF THE SDRIF (SCHÉMA D'ORGANISATION DE LA RÉGION PARISIENNE) OF PAUL MAYMONT, 1963-1966.

Other possible recollections of Gloeden can be found in the diagrammatic sketch contained inside the *SDRIF* (*schéma d'organisation de la région parisienne*) of 1963-1966 by Paul Maymont, which seems almost like a caricature. On the other hand, the diagrams of the Metabolists reassemble the cellular concept in new, fresher ways.

Before delving into the relationship between nuclearity and the archipelago, I would like to quickly examine the persistence in urbanism of concepts hailing from biology: of life, life cycle, and survival (ROBERT-DEMONTROND 2005). Baumeister in 1876 speaks in his manual[33] of the city's *natural* tendency towards growth, and Hegemann imagines the plan as a device that permits comprehension and achievement of the *natural* development of the city, in opposition to forces of speculation and the arrogance of absolute power, which disrupt the *natural* tendencies[34] of urban growth. Martin Mächler illustrates his plan in 1920 for Berlin in these terms: "An architectural work not in the sense of single constructions, but in the sense of a unified architectural creation, in which the single construction is a cell of a large, suitably articulated architectural structure, and in which this structure constitutes, in turn, a vital organism inside the great communitarian cellular tissue"[35]. Park describes the city in terms of natural areas, portions characterized by social or functional homogeneity, while the diagrams of Eliel Saarinen, used two decades later in his text *The City* (SAARINEN 1943), show the permeability of urban planning to imagery from cellular biology, and the resulting conviction that the reality of the urban slum can be understood by observing the degeneration of a tissue of diseased cells.

The theme of all the diagrams that take the cellular analogy as their starting point is the difficulty of conceptualization and representation of the relations between the individual and the group, of the forms they assume or can contribute to generate. This is a crucial question for urban planning and architecture. From the idea of social organization as a body, to that of the health of the social body, to the concept of urban evolution, we find ourselves immersed in the strongest system of analogies, similes, and metaphors that has constructed modern thinking about the city, considerably more influential than the analogy with the human body[36].

The Darwinian concept of evolution is the other great basis for all the theories of growth and transformation that introduce the temporal dimension in the design of the city. This emphasizes the capacity of species adaptation and the process of selection as the success of random variations that turn out to be useful for those same species. Shifted into the field of urban design, evolutionism and its more recent permutations in neo-evolutionism inspire categories of interpretation and design that move from the idea of death and life of metropolises[37], to that of "selective accumulation" or of attention to infinite individual variations, some of which introduce advantageous mutations for a community or a territory. Finally, in the evolutionist theories, the property of adaptation is joined by that of incremental mutation that has driven many recent research projects on the transformations of the territories of European sprawl.

Evolutionist and organic models share certain aspects, but there are also fundamental differences worth pointing out: the former interprets changes in the society in an evolutionary way, but does not suppose – unlike the latter – that society or the city should behave like living organisms. The organic analogy contains a systemic idea that is often taken, also in the contemporary debate, beyond the limits of the reasonable and beyond the proof of experience.

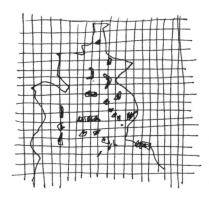

REDESIGN OF A DRAWING FROM DIE STADT IN DER STADT, BERLIN DAS GRÜNE STADTARCHIPEL, 1977, BY O.M. UNGERS.

To get back to the concrete territory, the cellular hypothesis of Gloeden seems to have been achieved in many European territories, for example in the city-territory of the Ruhr, in Holland, or in Denmark. Though the density is lower, the void is truly that equipped, continuous space described by the diagrams. The nuclei are varied, so much so that today they are no longer interpreted as cells, but as fragments; this is how Thomas Sieverts indicates them in *Zwischenstadt* (1997). In recent decades, a new interpretation and new problems have been superimposed on the quiet, reassuring vision of the cellular tissue – the theme of lost unity and of the relations among distant, diverse, and separated objects.

The idea of the archipelago, instead of the tissue of cells, investigates the relationships among fragments, expression of the multiple and of the distance between irreducibly different

Jérôme Chenal

THE WEST AFRICAN CITY

URBAN SPACE AND MODELS OF URBAN PLANNING

Rapid growth, unmanageable cities, urban crisis, macrocephali... The cities of west Africa are no longer 'plannable' – at least not using traditional urban development tools.

Without negating the importance of participatory processes in city creation, it nonetheless seems crucial to return to city plans and models, to what cities convey, and how they are built. But to understand the city in all its depth and richness, we must also hit the streets.

The West African City proposes a dual perspective. At the urban scale, it analyses historical trajectories, spatial development, and urban planning documents to highlight the major trends beyond the plans. At the second level – that of public space – the street is discussed as the city's lifeblood.

By innovating approaches and testing new methods, *The West African City* offers an unconventional look at Nouakchott, Dakar and Abidjan, the three study sites for this investigation. The city of today, in Africa or elsewhere, must re-examine its many social, economic, cultural, political, and spatial dimensions; for this, urban research has begun challenging its own methods.

This book is also the companion of Chenal's MOOC *African Cities*.

Author

Jérôme Chenal is a Doctor of Science (Ecole polytechnique fédérale de Lausanne, EPFL). After studying architecture at EPFL, he worked as an urban planner for the Urbaplan Agency, which specializes in urban development in Africa. He then joined the EPFL's Laboratory of Urban Sociology (LaSUR), where he wrote his thesis. He was a visiting researcher at University College London's (UCL) Development Planning Unit (DPU) for two years starting in 2010, before returning to the EPFL's School of Architecture, Civil and Environmental Engineering (ENAC), where he is heading the development of a curriculum in Urban System Engineering and has set up a MOOC about urban planning in African cities. He is currently the Secretary general of EPFL's Urban and Regional Planning Community (CEAT).

2014, 16,8×24 cm, 368 color pages, ISBN 978-2-940222-82-7

0 0.5km 1km 1.5km 2km 2.5km

NUCLEAR DESIGN AND THE ARCHI-
PELAGO CITY: A PORTION
OF THE RUHRGEBIET.

things. It seeks to establish not only the spatial but also the social characteristics of an aggregation of fragments. "The intelligence of the archipelago divides and separates", (Cacciari 1997), putting fragments into relation. It is the "fatigue" of a theory that leaves individualities intact but gathers them inside a space of coexistence – the sea – and of absence, the lost or never achieved unity – "Unfathomable and Unattainable", that philosophy indicates as "good". Islands are forced into dialogue because the "space of the archipelago is by nature resistant to subordination and hierarchical succession" (Cacciari 1997: 19-20). It is a space without center, in constant tension between the need for dialogue and its own individuality, its own interior.

In the declining Berlin of the 1970s, Ungers sees the process of emptying as the possible construction of a different principle of urban space (Ungers 1978), generated by projects that can best be interpreted as fragments, partial solutions for a specific site that move away from the logic of the comprehensive, rigid, inflexible plan (Ungers 1976). In the case of Berlin, the islands are "cities in the city" and the result of the erasure of portions of fabric that cannot be "re-qualified", or inserted anew in urban dynamics. The archipelago is the distance between things, but it is also their suppression, as well as the sudden appearance of gaps, the shrinking of the city, and its reduction caused by events that change its economics, demographics, and social characteristics.

Ungers is one of the first to grant visibility to a condition that differs from what had been approached by modern urban planning, and in which it was formed – in a context of progress, urging growth and development. The image of the green archipelago, now taken as an example in the research on the phenomenon of shrinking cities, or as a last barrier to settlement dispersion[38], has many precedents, as I have tried to demonstrate. From the plan of Scharoun for the reconstruction of Berlin (SOHN 2007) which organizes the city of stone described by Hegemann along the Spree Valley, drastically reducing its density and imagining habitat cells inside a sea of urban agriculture and green spaces; to the urban planning tradition of the *Stadtlandschaft*[39], introduced by the German geographer Siegfried Passarge around 1920; to the monist philosophy, whose Monist League was founded by Haeckel in 1906; all the way to the Monist scientist Raoul Heinrich Francé, admired by many modern architects; to the diagrams of Gloeden. Francé wrote about the communities of plants and animals as associations of cells, organs capable of more advanced tasks than those performed by the single individual. The organization of the village was praised for its capacity to multiply the qualities of its inhabitants, so that the cellular condition would permit extraordinary capacities to emerge[40].

Ungers frees the enclaves, the new islands that take on recognizability and singularity, leaving behind the "anonymity of the city" (UNGERS 1978). Between one fragment and the next, the green lagoon welcomes collective equipment, the space of large commerce, and recreation. The city is transformed into a union of fragments, of cells that forcefully evoke the organic metaphor and do not define a single unified image, but a "living collage" (UNGERS 1978).

The contemporary fragmentary and dispersed spatial condition, which is the result of a process of modernization in which individual, group, and society have rethought their way of living together, brings deposits of ideas to the surface, resting on profoundly altered contexts.

---

[1] For example, the volumes of Paolo Sica, *Storia dell'Urbanistica*, and in particular the volume *Il Novecento*.

[2] The theory of the isolated state (*Der Isolierte Staat*; THÜNEN 1826) is based on the premise of an isotropic territory, equally fertile and accessible in all directions with the city at its center, the location of the market. From this it follows that the value of the land diminishes in keeping with the distance from the center, and that around the city rings are formed with different uses of the

land, in relation to the distance. These diagrams, for example, give rise to the "milk radius" that has great importance in English planning in the 1900s.

3 "A single street 500 meters wide and long as is necessary; let this be well understood: as is necessary". In the newspaper *El Progreso*, Arturo Soria describes the linear city for the first time on March 6, 1882. See the anthology edited by George R. Collins and Carlos Flores in Soria y Mata 1968: 152-153 and HILBERSEIMER 1944: 68.

4 Different influences converge in the linear city, from the evolutionist theories of Darwin, Spencer, and the German biologist Haeckel, supporter of the theory of monads, who introduces the term "ecology".In the handwritten notes that accompany the copy belonging to Astengo, conserved at the Department of Urban Planning of the IUAV, of the *Principes de Géographie humaine* by Vidal de la Blache, Giovanni Astengo divides the header into two rectangles and places two titles: "*Schemi urbanistici*" and "*La città lineare*". At the title of the section, *Antenati della città lineare*, he indicates: 1/ valley of the Niger and 2/ plains of Tch'eng-tou, where the road 80 kilometers long lined by houses amazes the French mission, cited by Vidal de la Blache. The third reference, also from Vidal, is Apulia: "The coastal band extending from Barletta to Bari to Brindisi all the way to Lecce [...] organized according to a double parallel series of cities, one on the coast, the other about 10 kms, inland".

5 In the dense debate at the end of the century, the idea of a new relationship between city and country and of a return to nature crossed the work of many thinkers. Soria y Mata makes reference to Reclus and to the series of *The Evergreen* of Geddes, to which Reclus contributes with the text "La Cité du bon accord". See also (RECLUS 1866).

6 "Geddes" thinking machines were an idiosyncratic kind of diagram. They consisted of a number of divisions of sheet of paper through a process of folding. The matrix of fields would be filled with concepts which, through the process of folding, could be correlated in a tangible manner". DEHAENE 2002: 47; WELTER 2002.

7 "By descending from source to sea, we follow the development of civilisation from its simple origins to its complex resultants; nor can any elements of this be omitted... In short, then, it takes this whole region to make the city". As Helen Meller, who quotes the passage from Geddes, reminds us, the "valley section" had already been published in Geddes, P. , "Civics: as applied sociology", part 1, [in] *Sociological Papers*, Branford, V.V. (ed.), London, Macmillan (MELLER 1990).

8 HILBERSEIMER 1949: 89.

9 "In the big cities, civilization has undoubtedly left us a legacy whose elimination would require a great deal of time and effort. But they should and will be eliminated, though this will be a very laborious process" (ENGELS 2003: 356). The suppression of the city-country antagonism is the direct result of the evolution of the modes of production and the sole possibility of solving the problems of pollution and congestion of the city.

10 "The opposition between city and countryside is the most vulgar expression of the subjection of the personality to the division of labor that transforms the individual and reduces him to the status of an urban animal in one case and a rural animal in the other" (Marx). "Open recognition of the progress that can be attributed to big cities in the capitalist society does not prevent us from inserting in our program of action the elimination of the contradictions between city and countryside" (Lenin). The quotations from Marx and Lenin and the previous one from Engels are part of the reply of M. Ginzburg to Le Corbusier who had written to him before leaving Moscow, as he had been asked for an opinion on the results of the competition for the "Green City" of 1930. Together with expression of admiration for the European master, Ginzburg asserts the exceptional character of the Soviet situation and the originality of the project regarding the socialist city. See the two letters in Appendix 6 of the italian edition, KOPP 1970: 252-254.

11 Hans Ludwig Sierks was a traffic engineer and urban planner, author of *Grundriss des sicheren, reichen, ruhigen stadt* (1929). Sierks proposes the residential superblock schematized by Hilberseimer, and, like Alker Tripp, he points to the need to modify, if not to definitively abandon, the "block system" to solve the problem of dangerous intersections. The sector is therefore the assemblage of multiple blocks inside an area of about two square miles, bordered by streets of a higher order (HILBERSEIMER 1944: 104). The superblock conceptualized by Sierks is interpreted by Hilberseimer as a new settlement unit, a new city element. This is the "brick" from which to begin

to rebuild an "organic structure for the community life of the people" (HILBERSEIMER 1944: 113). Hans Ludwig Sierks (1877-1945) was *Stadtbaurat* of Dresden from 1919 to 1924. He took part in the "Proletarischen Hundertschaften" movement in 1944, and in the failed attempt to assassinate Hitler on July 20. He was arrested and executed in 1945 in Berlin. He involved his friend Erich Gloeden (see the next paragraph) in the plot, asking him to hide the general Fritz Lindemann.

[12] With Peter Friedrich, Hilberseimer corresponds from the 1930s to the mid-1960s (see *Ryerson and Burnham Archives, Ryerson and Burnham Libraries – The Art Institute of Chicago*). The archive contains writings and articles by Friedrich on the advantages of the linear form.

[13] HILBERSEIMER 1944: 126.

[14] Arminius (Adelheid Dohna Poninski), *Die Grossstädte in ihrer Wohnungsnot und die Grundlagen einer durchgreifenden Abhilfe*, 1874. Also see the plan for London, ahead of its time, by John Claudius Loudon in 1829 (http://www.londonlandscape.gre.ac.uk/1829.htm) and PANZINI 1993.

[15] Edward Gibbon Wakefield (1796-1862) played an important role in projects of colonization of southern Australia starting in 1831, and later of New Zealand. He wrote *A Letter from Sydney* (1829) and *A View of the Art of Colonization* (1849), setting forth his theories on the need for systematic colonization. The law launching the creation of colonies in South Australia is from 1834; the British government forms a Board of Colonisation Commissioners that provides Colonel Light with precise indications on the planning of new cities, regarding (for example) the need for wide streets, squares, and public promenades. The Board urges that examples be sought in America and Canada where many new cities, such as the Philadelphia of William Penn founded in the second half of the 17th century, were built on a grid, and some of them, like Savannah, founded in 1733, and other cities in Georgia, were surrounded by a circular parkland. If the grid expressed the idea of an egalitarian space (each colonist was assigned a parcel and an agricultural lot outside the city of similar size), the void around the city was also an element of defense. See the website of the Australian Heritage Database, and for the European roots or those connected to the Spanish colonization of American cities (REPS 1965).

[16] He writes *A General Plan for Laying Out Towns and Townships, on the New-Acquired Lands in the East Indies, America, or Elsewhere* in 1794.

[17] Victor Aimé Huber, cited by Hegemann (HEGEMANN 1975: 226). Physician, professor of languages and a social thinker, he theorizes the *innere Ansiedlung*.

[18] A copy is found at the Centre Canadien d'Architecture, with a dedication to Ruth Maria Baumann, written by the author.

[19] I have not been able to further reconstruct the biography of Gloeden the "perfekter Assimilant", as he said of himself, but we can at least add that from 1936, after the death of his father, he returned to the Jewish faith and embraced Zionism, though participated as an architect in the Nazi organization Todt.

[20] The first electrified line is opened in 1924; the term *Stadtschnellbahn* is introduced in Berlin in 1930.

[21] See chap. 2, paragraph I: "The total elimination of automobile traffic inside the city".

[22] GLOEDEN 1923: chap. 2, paragraph IV.

[23] GLOEDEN 1923: chap. 2, paragraph V.

[24] GLOEDEN 1923: chap. 2, paragraph III.

[25] Christaller also supports Nazism and uses his theory of localization to construct a plan for the reorganization of Poland.

[26] A copy from Astengo's library conserved at the IUAV, University of Venice.

[27] ROSSI 1987: see the introductory essay. Antonio Tosi writes in the essay *Verso un'analisi comparativa delle città*: "The conceptualization of the city and the construction of urban theories become the prerogative of the social sciences". Referring to the Chicago school, he underlines the influence of biological ecology and social Darwinism and the separation into two levels of analysis, cultural and biotical, as a mode of transforming a set of empirical data regarding the American city into a general theory.

[28] VV.AA, *Idee per la città comunista*: Giancarlo De Carlo translates it into Italian and publishes it in 1968. In 1971 De Carlo's version is translated in English as *The Ideal Communist City*.

[29] VV.AA., *The Ideal Communist City*, p. 97

30 De Carlo's note inside the front cover that introduces the series *Struttura e forma urbana* conveys an idea of the cultural atmosphere in which the book was translated and published in Italy. 1968 is also the year in which De Carlo curates the Milan Triennale, on the greatest exponents of architectural structuralism, and the exhibition never opens due to student protests.

31 VV.AA., *The Ideal Communist City*, pp. 101, introduction to the chapter, Structure of the "Urban Environment".

32 VV.AA., *The Ideal Communist City,* pp. 112; 114.

33 Baumeister, R., *Stadterweiterungen in Tehnischer, Baupolizeilicher und Wirtschaftlicher Beziehung* (Town extensions: their links with technical and economic concerns and with building regulations), Berlin, Ernst Korn, 1876.

34 In her introduction to *La Berlino di pietra* (Stone Berlin) of Hegemann (1930), Donatellà Calabi examines the idea of natural growth of the city as an objective to be pursued, a theme that returns in German thinking starting from the second half of the 20th century; see HEGEMANN 1975. In PICCINATO 1974, there is an anthology of the manual of Reinhard Baumeister.

35 Martin Mächler, "Die Deutsche Weltstadt-Aufgabe" (the functions of the German metropolis) in Taut, B., 1963, *1920-1922 Frülicht, Eine Folge für die Verwirklichung des neuen Baugedankens*: "Eine architektonische Zeit nicht im Sinne der baulichen Einzelschöpfung, sondern eines baulichen Gesamtschaffens, das das Einzelwerk zur Zelle einer größeren architektonisch richtig gegliederten Gemeinschaft gestaltet und das diese Gemeinschaft wiederum as lebenskräftigen Organismus, in das große gemeinschaftliche Zellgewebe des Staates einfügt", pp. 188.

36 See, among others: SECCHI 1984 and CAVALLETTI 2005.

37 Just consider Lewis Mumford and Jane Jacobs.

38 See, for example, in the present French context, the use of the figure of the archipelago (VELTZ 1996) also as a constructive metaphor of plans on a territorial scale (CHAPUIS 2003).

39 "Stadtlandschaft" is an expression coined around 192, by the geographer Siegfried Passarge (1867-1958) (*Die Grundlagen der Landschaftskunde*, Hamburg 1920/21), also the author of *Stadtlandschaften der Erde* (Hamburg 1930). Thanks to Thomas Sieverts for the reference. Also see MANTZIARAS 2008.

40 See BOTAR 1998.

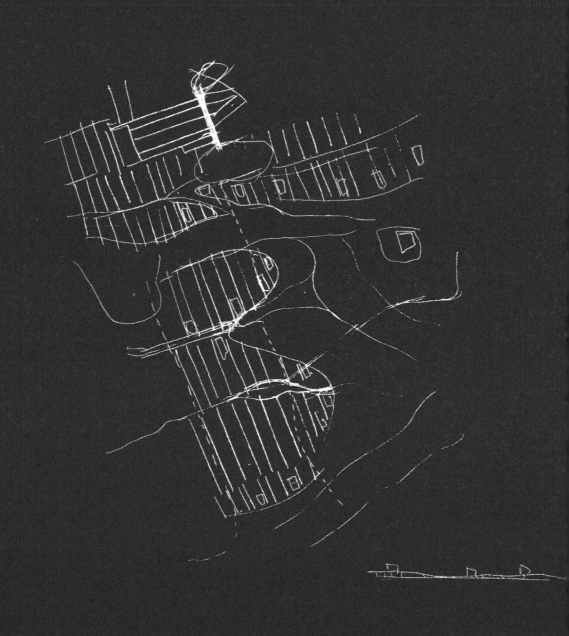

Figures of the void

Green ribbon.   If the green belt becomes the backdrop and setting in which the new urban spatial character is achieved, the ribbon, canal, and green corridor introduce a concept of 'void' that detaches itself from the background and becomes a figure in its own right. This takes on explicit theoretical consistency in the texts of *Landscape Ecology* only in the second half of the 1900s[1], but it emerges as a reference of urban and territorial projects at least one century earlier, as a result of careful interpretation of the characteristics of place, from practices linked to landscape, to the promenade, and to the outdoor movement in general. Some short notes on its formation shed light on the connection between the concept of the *green corridor* and that of a system of open spaces during the same period.

In 1859, the special committee for the creation of the parks of Boston is appointed by the Boston City Council to organize a competition for the Boston Public Garden. The committee indicates that Boston has neither the size nor the configuration suitable for a large park. Instead, it recommends using the small portions of open space that have remained intact. In 1869, after the Civil War, a new committee decides on the purchase of a large area of land for a park, or for a series of smaller parks[2]. H.W.S Cleveland, a landscape designer, writes in *The Public Grounds of Chicago* that Boston does not need a large park, but "a system of improvements over the surrounding country […]"[3] that also includes the planting of trees along streets. Also in 1869, Uriel H. Crocker, a Bostonian notary public, writes a letter to the committee and to the newspapers, accompanied by a plan showing a system of linear parks. Utilizing the natural characteristics of the landscape, the parks connect the city to the countryside; the system is proposed as an original form of the "Central Park" without the need to imitate its model[4]. In 1870, the Park Act is passed which is supposed to institute a metropolitan commission, but it fails to do so. In the course of the extensive debate a young man stands out, Ernest Bowditch, who proposes connecting the various parks with a boulevard, probably based on the ideas of the same era of Robert Morris Copeland. In the meantime, Olmsted is repeatedly asked to offer his opinion. He gives a lecture in Boston in 1870, entitled "Public Parks and the enlargement of towns", in which he defines the park as "a simple, broad, open space of clean greensward"[5]. In 1876, the Boston Park Commissioners produce a long description and map of the Boston park system.

REDESIGN OF THE
METROPOLITAN PARK SYSTEM
OF BOSTON BY URIEL H.
CROCKER, 1869, TAKEN
FROM ZAITZEVSKY, C., 1982,
FREDERICK LAW OLMSTED AND
THE BOSTON PARK SYSTEM.

The concept is clarified over time, through projects often not built or, in some ways, not implemented; the green corridor, in this case, is not the application of a theory other than that professed, for example, by Olmsted. He is an attentive interpreter of the physical characteristics of the landscape, in keeping with the specificities of a city, with an eye on minimizing costs of construction and maintenance, and on the function of collecting and purifying water. In 1869, the plan of Olmsted and Vaux for Riverside, Illinois, is crossed by a green channel, a park organized along the river, to which other, smaller open spaces could be attached. The project calls for separation of the movements connected with free time and those connected with work. A "pleasure parkway" (FABOS, MILDE, WEINMAYR 1968: 50) six miles long would connect Riverside to Chicago. The promenade is never built, but the space along the Des Plaines River becomes a "public common".

In Boston, these first attempts find concrete form in the idea of a system of green areas, intended first of all to solve health issues connected with the presence of swamplands (the Back Bay Fens) into which the city's sewers are discharged, and with the risk of flooding from the two rivers that converge on the Back Bay, the Muddy River, and the Stony Brook. In 1878, Olmsted designs a linear park that drains the area and transforms the marshy zones into a lamination basin for periods of heavy rain, while constructing sewer pipelines and tide control systems of the Back Bay (FABOS, MILDE, WEINMAYR 1968: 58). In the description of the green ribbon within a presentation made to the Board of Commissioners[6], Olmsted emphasizes the importance of the connections between one park and the next, which make it possible to get beyond the idea of distinct, independent green episodes (also an expression of strong local sentiment in the case of villages or districts gradually being absorbed by the metropolitan area), and clarifies the possibility of constructing a "system" by exploiting the complementarity and differences between the parks of the "necklace"[7].

Sylvester Baxter, a journalist and disciple of Edward Bellamy, author of *Looking Backward, 2000-1887*, published in 1888[8], becomes the leading defender of Olmsted's projects for Boston. An expert on German urban planning and a supporter of the *Metropolitan Park System* for which he is the secretary of the commission in 1892-1893, Baxter asserts that in the design of the parks, in city planning and regional planning, one finds "the most immediate application of Nationalist theories" (FEIN 1972: 60-61). With respect to Olmsted's experience, the idea of the city has changed, and the scale of the village and its community gives way to the design of the large, regular city.

The text *Town Planning*, published in 1916 by John Nolen, reports on the spread of the idea of the park system, i.e. a set of interconnected areas. In his list, he distinguishes between two groups: "scattered facilities" like playgrounds or neighborhood parks, and "connected facilities" like urban parks, reserves, and forests with "parkways"[9]. This distinction is enlightening because it clarifies that the idea of the system, from the start, is linked to that of physical interconnection of the various spaces. One surprising experiment is the plan for San Francisco devised by Burnham, a friend of Olmsted, which — if observed without focusing on its obsession with diagonals — reveals a structure of open and public space composed not only of streets, squares, and highways, but also of "park connections". The plan, begun in 1904, is of importance not so much from the viewpoint of the concepts proposed, as for the descriptive approach to the design of the city, which I will discuss in the second part of this book.

The Jansen plan, prepared for the competition for greater Berlin in 1910 with the motto "In den Grenzen der Möglichkeit" ("within the limits of possibility"), also contains a detailed concept of open space, in which a set of green spaces organized as a ring are attached to the large river corridor. The space-corridor not only represents the continuity between the green areas, but also imposes the theme of transition, due to the absence of clear borderlines between one space and the next. In the map "*Wald und Wiese Gürtel um Berlin, mit radialen Verbindungen*", the colors take on shadings that indicate the passage from the forest to cultivated areas to parks, the transition between different ecosystems, and the appearance — though not yet explicit — of the concept of the "ecotone".

Jansen's schema becomes an important reference for the "General Plan of Open Spaces" (*Generalfreiflächenplan*) of Martin Wagner, who works on the master plan of Berlin from 1925, the year in which he is appointed *Stadtbaurat*, until 1933. Jansen's drawings form the basis of the plan for the green zones of the city of 1929[10]. Martin Wagner was familiar with the park systems of the United States, analyzed in his text from 1929 on the American city and its problems (WAGNER 1929). In his most famous diagram, the open spaces are divided into two major categories, restricted spaces and agricultural territory. The former are clearly imagined inside a hypothesis of continuity, the same one that has guided, until recent years, territorial and urban design. The penetrators and corridors emerge, a set of green spaces investigated by Wagner since 1914 when he prepares a doctoral thesis on the theme of healthful greenery in the open space of the large metropolis. The penetrators push into the city and attach some of the isolated green spaces represented by Jansen to the overall system; the corridors contain the new city portions (*siedlungen*) inside a network and on different scales. With respect to Jansen's map, the idea of transition, of fluid passage from one state to another, is lost. The contours are clear, the legend is simplified and reduced.

From 1920 to 1923, Fritz Schumacher prepares a greenery plan inserted in the reflections on "greater Cologne" in which, together with multifunctional parks and wooded reserves, "green canals" appear that connect the parks to the city and to the large external spaces.

REDESIGN OF THE COMPETITION PROJECT FOR GREATER BERLIN BY H. JANSEN, "WALD UND WIESE GÜRTEL UM BERLIN, MIT RADIALEN VERBINDUNGEN", 1910.

REDESIGN OF THE "KARTE VON FREIFLÄCHEN IN DER STADT BERLIN UND UMGEBUNG", OF W. KOEPPEN AND M. WAGNER, BERLIN, 1929.

REDESIGN OF THE GREENERY SYSTEM OF F. SCHUMACHER FOR COLOGNE IN THE 1920s.

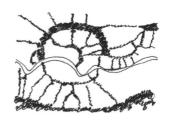

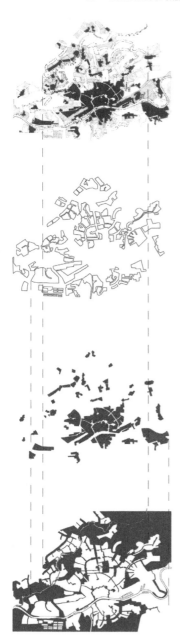

Strongly influenced by Olmsted and Sitte, Schumacher speaks of a green belt and a system of parks. Two *Grüngürtel* elements are built on the old fortifications to be demolished following the peace treaty of the First World War and part of the demilitarization of the Rhineland region. Completed in the 1930s, the two green belts and green corridors still structure the space of the city of Cologne today.

Regarding the experience of Ernst May in Frankfurt, from the same period as those of Berlin and Cologne, observers and historians point to the renewed syntax and concepts of housing design, while generally criticizing the lack of a wider-ranging vision. This critique has often been extended to the entire contribution of the Modern Movement, for its supposed incapacity to structure the city on a large scale. Trying to redraw the completed projects and the planning schema produced by May in Frankfurt (the zoning plan of 1930), several doubts arise about this interpretation. Observing the sequence of interventions along the Nidda River, the idea of the medium in which to position cells seems to emerge anew. Even if the *Trabantenprinzip* is of English and Howardian origin, the open space of the big city is similar to the diagrams of Gloeden. The fluvial corridors are part of a larger surface in which the different types of movement, also from an ecological viewpoint, can take place. The *grüne Gürtel* is increasingly similar to a network of parks, promenades, wetlands, urban vegetable gardens, and cemeteries that are connected to the city center. That center separates, but does not create satellites; rather, green spaces and waterways delineate a new urban continuum.

With different terms in different languages, the theme of the green corridor, of the "greenway", is also seen in the later plan of Louis Kahn for Philadelphia, based on the suggestion of Edmund Bacon, chairman of the Philadelphia Planning Commission, who experimented with it immediately after World War II (SMITH MORRIS 1965; BACON 1967). The reference in this case is to Geddes and his principle of "conservative surgery" proposed to approach the re-design of vast decayed areas in India. The approach is descriptive and "It requires a long and patient study. The work cannot be done in the office with ruler and parallels, for the plan must be sketched out on the spot, after wearying hours of perambulation"[11]. The "intimate knowledge of the individual characteristics of an area"[12] does not stop with the careful surveying of trees, pavements, and existing furnishings, but also calls for the commitment of citizens to fill urban space with their own individual expressions.

As a concept of organization of space and landscape, the design of the greenway is a moment of individuation and underscoring of the physical and social unique-

DECONSTRUCTION OF THE PLAN OF LAND USE OF THE CITY OF FRANKFURT BY E. MAY, 1930. FROM THE BOTTOM: OPEN SPACE AND RIVERS; EXISTING EDIFICATION; NEW EXPANSIONS; THE PLAN.

ness of space. The selected vegetation has to express the characteristics of the land. Its shifting levels are interpreted in design terms, while axes correspond to the lines of desire suggested by spontaneous pedestrian movements. Thus every greenway is an autonomous expression of the uniqueness of places. It resists standardization and mass production, and it is continuous in space and flexible in time, with no center and no outskirts. The concept is read in rather emphatic terms as "the beginning of new space-time urban design concepts" (SMITH MORRIS 1965: 32). The result is actually a far cry from the plan of San Francisco by Burnham, despite the asserted distinction between a "static" project and one capable of grasping movement[13].

The points I have briefly illustrated are useful for several reasons. The first is that modern urban planning and, more in particular, the Modern Movement have always been unjustly accused of not having produced concepts capable of structuring the new urban and territorial scale. In the view of many historians of urbanism and the city, the *siedlung*, the housing estate, is the only scale approached and resolved by modern urbanism. The reflections I have outlined point to a different interpretation: what the Modern Movement can be accused of is of not always having been in the avant-garde, but of having absorbed ideas and experiments that already had a long history. This, however, is another matter.
The second reason for interest is that the concept of the green network appears first as a figure of urbanism and only later as a functional concept related to theories of landscape ecology. The ambiguities of the green belt concept, the important differences between it and the *grüne Gürtel* – its supposed rigidity of form and dimensions, the lack of an explicit concept of the green network and its synthetic representation – have perhaps delayed and limited the comprehension of the conceptual innovation contained in the modern design of the city and the territory.

Spatial structures and procedural diagrams

In the 1960s, the arrival of systems theories represents the framework in which concepts and diagrams are developed that approach the theme of complexity and coexistence on different scales. They approach multi-dimensional, interdisciplinary issues in a context of uncertainty; they have to provide "generative results, that is to allow for and support the emergence of second-order effects"[14] (LOBSINGER 2000). Some of the diagrams of Cedric Price show functional overlays, programmatic matrices of architectures and settlements that seem to simultaneously present themselves as pure support (in the Potteries Thinkbelt project, for example) and as pure space, without the problem of having to supply society with symbols through architecture. The harsh criticism of this approach in the years immediately to follow marks a change of perspective that also corresponds to the decline of the diagram and of conceptualization as an activity of design and its cognition.

The following paragraphs illustrate the rise, during the 1960s, of concepts and diagrams that are no longer interested in producing a form. They set out to clarify a structure that can sustain different formal interpretations; diagram and concepts attempt to represent the process and the conditions in which multiple forms can be generated over time.

Semilattice.   In 1965, Christopher Alexander publishes *A city is not a tree*, an essay in which diagrams of two types of structures are contrasted: the tree structure, which in Alexander's view is the matrix of all modern urban design, and the "semilattice" structure in which the spontaneous, undesigned city is represented – along with the ambitions of a theorist who wants to get beyond the fundamental concepts of the modern project. The two structures offer different interpretations of the ways in which a large collection of many small systems can generate a larger, more complex system[15]. The first – the tree structure – reduces the complexity but oversimplifies the relations among the parts, restricting them to dependence or mere juxtaposition. Given a set of objects, either one is totally contained in the other, or it is totally separated. In a tree structure, this "means that within this structure, no piece of any unit is ever connected to other units, except through the medium of that unit as a whole"[16]. Observing the best-known projects of modern urban planning, from the New Towns of Clarence Stein to Brasilia and Chandigarh, Alexander points out that they share the idea of the tree structure. The best example of the fascination with this structure can be found, according to Alexander, in Hilberseimer's text *The Nature of Cities*, where the author refers to the military camp as the archetype of the city. Not only the Roman city but also the modern city can be recognized in the idea of hierarchical order and discipline typical of the tree structure, and a shopping center can be designed starting with the archetype of the military camp.

LONDON AS A TREE STRUCTURE IN THE PLAN OF ABERCROMBIE, FROM C. ALEXANDER, A CITY IS NOT A TREE, 1965.

The second structure, the "semilattice", is a collection of objects partially organized in a hierarchical sense; it is an interpretation of the lattice, which has countless definitions in different scientific fields. Alexander uses it to introduce the theme of overlapping in the conception of urban design. The variety introduced by the possibility of overlaying objects that go into the whole, or horizontal and not just vertical relationships, multiples the richness and density of the collection.

STEM: INTERPRETATIVE REDESIGN OF THE BOCHUM UNIVERSITY OF CANDILIS, JOSIC, WOODS, 1963.

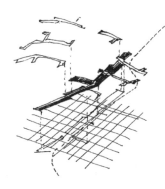

Stem, Web.   In the years prior to Alexander's text, certain attempts to get away from the idea of functional separation and the tree structure were expressed in a large series of diagrams produced by a part of Team X, in particular by Shadrach Woods and the Smithsons. The linear structure is again proposed as the most suitable for an open society (CANDILIS, JOSIC, WOODS 1965). The "stem" concept described by Woods in the articles of the early 1960s (WOODS 1960; 1961) makes reference to a "linear tracé operational as conceptual instrument of urban planning"[17] that puts the theme of the street back at the center of thinking on urban planning – also seen in an abstract sense, rather than as a reproduction of the spatial character of the ancient city. The line, in fact, has no form, no dimension. Woods starts with a radical critique of the project conceived as a *plan masse* and counter it with a topological order, a support for the location of different functions and the interaction based on the "network" of paths inside the fabric of the ancient city (Tzonis, Lefaivre 1999). The idea of redetermining multiple points of access to the dwelling, of rediscovering a link between community and individual, of organizing activities and services along a street crossed by public transport (a "street" and not a "road") to which the lodgings should be attached, ("plugged-in") remains, nevertheless, anchored to the modern project of the separation of flows. The linear arrangement of the system ensures its ability to grow in time and to adapt to different types of housing and density.

The "web" is produced by the interconnection of multiple stems. Like them, it is an open, evolving and a-centric system (Woods 1962; Candilis, Josic, Woods 1965), which can develop in space, not just on one plane. It is a product of relationships and associations. The stem and web express a profound critique regarding both the planivolumetric aspects of architecture – with the pursuit of allegories and symbols through the play of volumes and the pursuit of a new, suspect monumentality – and the planning of functional purposes. The idea of continuity necessary for a city and an open, growing society is the pursuit of isotropic conditions, equal in all directions. No part should remain isolated or be subject "to an a priori over-densification" (Candilis, Josic, Woods 1965: 188)[18]. With this same approach, Alison and Peter Smithson inserted the street among the categories that defined the *Urban Re-Identification Grid* for the CIAM of 1953 at Aix-en-Provence.

The richness of the relationship between street and block is again emphasized, during the same years, in the studies conducted on the fabric of Paris by Chombart de Lauwe (Chombart de Lauwe, Antoine, Bertin, Couvreur, Gauthier 1952); by Gutkind, who analyzes traces as "elements that structure the spatial practices of the urban realm"[19] in his text *Expanding Environment* of 1953; and by the critical interpretations made by Henri Lefebvre of the *grands ensembles* being constructed in France (Lefebvre 1968)[20]. The study of the logic of construction and growth of the European city conducted by Candilis, Josic, and Woods, admirers of Chombart and Lefebvre, had the aim of revealing the durable, permanent elements of the city, which, precisely due to their permanence, can make change possible.

Procedural diagrams / 1.   The competition for Parc de la Villette, at the start of the 1980s, was an opportunity to compare the ideas of the city that were starting to clash with the theme of the fragment and the discontinuity of space. Some of the projects submitted for the competition contrast new urban characters and forms with proposals that bring out unity and detail, while others assume the fragmentation of the contemporary condition, representing it through a project conceptual-

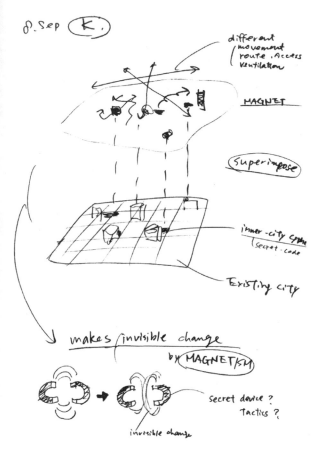

ℓ.Sep Ⓚ

different
movement
route . Access
Ventilation

MAGNET

Superimpose

inner-city Spa
(secret-code

Existing city

makes invisible change
by MAGNET(s)

secret device ?
Tactics ?

invisible change

C. PRICE, MAGNET "INVISIBLE
CHANGE", PRICE ARCHIVE (DR
2006:0018:008), CCA.

ized in terms of multiplicity and fragments, in the uncertain, non-finished space of action. To think about these terms and the themes they suggest, I will use some of the diagrams of Price, Koolhaas, and Tschumi for the competition for Parc de la Villette.

The case is particularly apt because it is a park, and the park has always played the role of a laboratory in the history of the city (Laugier wrote: "let the design of our parks serve as the plan for our towns"[21]), the place of experimentation and development of new urban theories. It has often not only represented a metaphor of the city which it reproduced and to which it alluded, but also brought innovation to the vocabulary of spaces, proposing new sequences. In the case of Parc de la Villette, the reflections and discussions that emerged from the competition, which concluded with the implementation of the project by Bernard Tschumi, address not only how to design an innovative space, but also more general questions that impact the contemporary city and the modes of its design and construction.

Parc de la Villette occupies an area of 50 hectares in the northeastern part of Paris, in what until 1974 had been the location of the slaughterhouses of Paris; in 1982 Jack Lang, the minister of culture, announced an international competition for the design of the new park. The modifications that took place over time in this part of Paris had formed an area at the edge of the central part, but connected to it by the Canale de l'Ourcq that opens into the large basin of the Villette closed off by Place Stalingrad (redesigned a few years after the competition by Bernard Huet) and by the Barrières, the old customs barrier of Ledoux. The flat, slightly hollowed area is located between two higher zones, the Butte de Montmartre to the northwest and the Butte Chaumont to the south, and bordered to the east by the Boulevard Périphérique which took the place of the fortifications of Thiers (1859). An area of transition between center and *banlieue*, the Villette is inserted in a fabric of heterogeneous, suburban characteristics. In the years prior to the competition, interventions of various kinds were done inside the area, leading to the construction of residential buildings, of the *Grande Salle* (an enormous building: 270 x 110 x 40 m), later transformed into the Museum of Science, Technology and Industry, and to the restoration of the *Grande Halle* as a place for entertainment.

The viewpoint of the EPPV (*Etablissement Public du Parc de la Villette*) that outlines the competition guidelines is of particular interest. It outlines an evolution of
Parisian parks from the 17th century: places of centrality, of gathering, with urban
facilities, then green Haussmannian spaces, "antidotes to pollution and the lack
of comfort of the city"[22], primarily frequented by children, mothers, and senior citizens. Between the postwar period and the 1980s, a weakening of the meaning of
the urban park is observed, as it is transformed into a quantitative factor and mere
mode of accompaniment to the buildings. The commission notes that "the parks
are not suited to our times" and that "symbols of the present should be introduced
in the contemporary park"[23], transforming it into a new cultural tool.
Among the various strategies proposed by the competitors, based on historical
continuity, functional relations, or flexibility, some make reference to formal tactics driven by fragmentation. In certain cases, it can be seen at all project levels.
The fragment is the relic of the existing and the decontextualized, capable of shedding light on complexity and possible inconsistencies. The architecture permits a
system of contradictions inside which the relationship between signifier and sig-
nified is not perceived as definitive. The representative and rhetorical dimension
of the project is important; the contemporary condition to be represented has plu-
ralism as its main symbolic and cultural reference, with new ways of experiencing
urban space[24]. Alongside the more famous projects such as those of Tschumi and
Koolhaas, which I will discuss briefly, I would like to examine some diagrammatic
sketches of Cedric Price, accompanied by the project description.

"A lung for the city. A twenty-four hour workshop where all can extend their knowledge and delight in learning"[25]. The project by Cedric Price for Parc de la Villette has
been practically overlooked: it is not even taken into consideration by Lodewijk Baljon who conducts in-depth analysis on many of the competition projects, despite
the paradoxical fact that certain commentators indicate the influence of Price (the
Fun Palace) on the projects for Parc de la Villette of Tschumi and Koolhaas[26].
Some years earlier Price had published, together with Hall, Banham, and Barker,
"Non-Plan" (BANHAM, BARKER, HALL, PRICE 1969), a manifesto against the idea of planning. In the years after the competition, he proposed a manifesto for London and
the metropolis, with the study "Magnet". In "Non-Plan", Price and the other authors
reinterpret urban planning as the expression of fashions and the producer — in
the best of cases — of secondary non-planned effects that are often more interesting than the initial project. In the article they propose a rigorously observable
experiment of "non-planning" in the heterogeneous places of change. The desire
is to *know* (in italics in the text) rather than to *impose*, to reveal the contemporary urban planning "style". Using descriptions, maps, and views, the four authors
develop a kind of trend scenario for certain English territories. These examples
convince them of the potentialities implied in practices of use of the territory and
of the capacity of such practices to create new types of spaces and new coexistences.

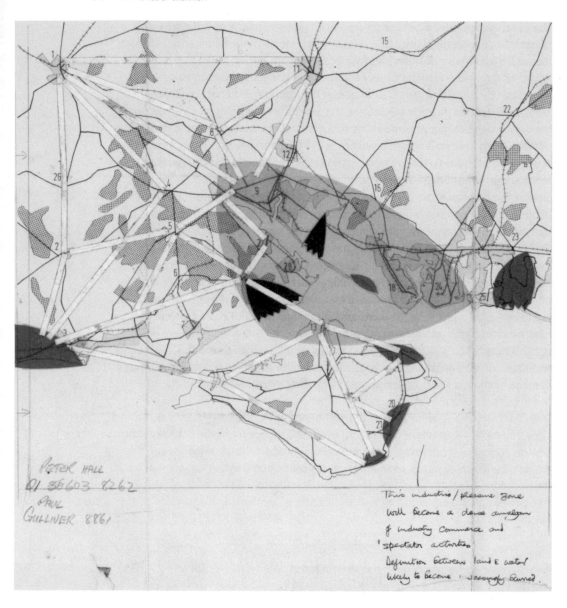

While the tone of "Non-Plan" is reminiscent of another important article on the new configuration and dimension of the city, "La città territorio" by Piccinato, Quilici, and Tafuri, the images, sketches, and diagrams add an ulterior level of comprehension and density. Before delving into the project for Parc de la Villette, I would like to look at the schema drawn for the "Montagu Country", named for the English aristocrat, the third territory of "non-planning" after "Lawrence Country", in the East Midlands between Nottingham, Derby and Sheffield, and the "Constable Country" northeast of London.

The Montagu Country has to do with the region of Solent, the arm of the sea that separates the Isle of Wight from England, a complex estuary system also faced, further in, by the port of Southampton. According to the authors, the region is a curious "hodgepodge"[27] of heterogeneous elements — factories, refineries, growing universities, tides, ports. It is the start of a megalopolis.

In "Non-Plan" the schema of Montagu Country represents a "potential industries/pleasure zone", a new settlement configuration based on the possible coexistence between industrial and recreational use of the territory. In the article the schema appears in black and white, but in Price's archives conserved at the CCA, the drawing has colors and certain differences, first of all that of the title: "Nul-plan" instead of "Non-plan", with the caption "Existing constipated conditions and potentials". With this different title the exercise performed in "Non-Plan" is clarified as an attempt not so much to abolish planning as to construct the project of the city in a different way, having the capacity to read the territory as it is being transformed and to interpret its potentialities. In this case, the "plan zero" can also sustain an experiment on not just — or not so much — urban planning rules, as more in general on the rules of social coexistence. In the conclusions, for example, the authors emphasize that the problem is not to eliminate planning, but to avoid its abuse when this word is used to impose "certain physical arrangements, based on value judgments or prejudices" (BANHAM, BARKER, HALL, PRICE 1969: 442). This idea of planning is inspired by systems analysis, as "frameworks for decisions", where the gathering and availability of information is seen as interesting but not sufficient[28]. The three revolutions that mark this period — the cybernetic revolution, a transformation that is not only technological but also cultural; the revolution produced by mass affluence[29], which leads to the demand for more space around ourselves and our activities; and the youth pop-cultural revolution — radically alter expectations regarding inhabitable space, to which traditional planning, the expression of an 'old bourgeois culture', cannot respond. The territory has to be a place of decentralized innovations, of new spaces in relation to lifestyles that are different from the past, and the expression of new cultures.

My hypothesis is that the diagrammatic schemes of Price attempt to add something to the analysis and the suggestions laid out in the text. They explore, constructing a sort of "scenario zero", the potential deduced from the analysis of practices (more leisure time, more extensive use of the territory thanks to greater mobility, the pursuit of nature and open spaces, the space of work) and from the observation of a specific territory, the estuary region, marked by a special, always variable relationship between land and water. Potentialities and possibilities intertwine in the schema: the possibilities of scenic movement along streets; the pleasure of a provisional, light use of the territory; the potentialities of a new relationship between industry and pleasure around the estuary, where working activity and the practices of free time can mingle; and where new lifestyles, "a new kind of living for Britain", could emerge (BANHAM, BARKER, HALL, PRICE 1969: 440).

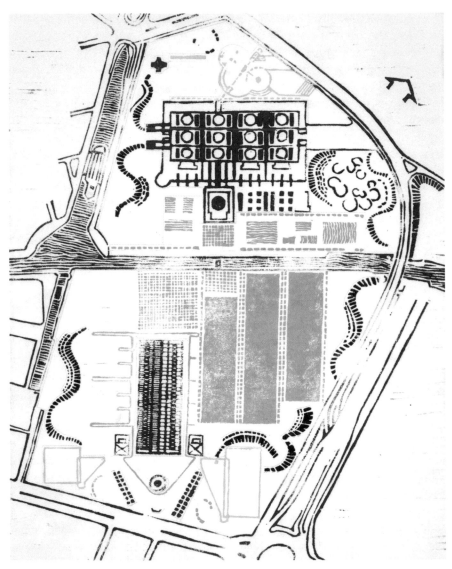

C. PRICE, NULL PLAN, PRICE
ARCHIVE (156 PARC), CCA.

In the Villette case, Price imagines the park as a place of production: of fruit, vegetables, fish, culture, research, music, and science. The innovation seems to arise in the conceptual space, even prior to the physical space. It has to do with the coexistence of things and people. The sketches show a park that is not totally occupied and designed, but is immediately ready for use. The worksite is already a park, a public place for events, leisure and relaxation, relationships. It is crossed by aerial paths, 20 meters wide and 180 meters long, self-propelled environments that can make slow movements, spanning the canal and the cultivated fields. These elements appear as "a clue to the fourth dimension: movement"[30]. These bridges move by means of horizontal hydraulic pistons and are places of enjoyment of the park for the public and the researchers; services and equipment are attached to them. The edges of the park are places of exchange with the rest of the city and the routes. The park is a "constructional toy", a social game of construction and exchange of types of knowledge. Each of its phases – building, modification, cultivation – is imagined as an opportunity for a public event, "a working exhibit"[31]. The exhibits are maintenance operations, good organization, the economy of the park, taking care of it, both static and dynamic elements, complex technology, and archaic features. The outfitted bridge is a tool to achieve all these aims: "a four-dimensional servant", useful for all the purposes, including that of agriculture.

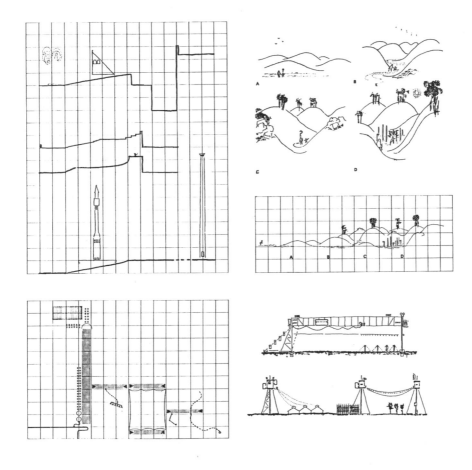

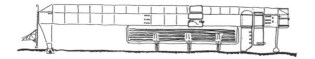

C. PRICE, COMPETITION PROJECT FOR PARC DE LA VILLETTE, 1980, PRICE ARCHIVE (156 PARC 1.3.3 REPORT CONCOURS PARC DE LA VILLETTE), SKETCHES AND PLAN, CCA.

C. PRICE, COMPETITION PROJECT
FOR PARC DE LA VILLETTE,
1980, PRICE ARCHIVE (156
PARC 1.3.3 REPORT CONCOURS
PARC DE LA VILLETTE), CCA.

linked gantries not only dispensing
their services to the areas they span...

... but also providing high level
public observation & access routes
at right-angles to the established
ground level routes.

The concept of the park is absolutely original and asserts that every public place is the result of work. The space is produced, and this transformation is an opportunity for the exchange and spread of knowledge. It is neither an educational park nor one solely of activities – as in all Price's works of architecture, irony wins out – but a place of overlapping roles. It is that "philosophy of enabling", spoken of by Royston Landau, a colleague of Price at the AA (PRICE 2003: 11), whose roots lie in the utilitarianism of Bentham. It is a market, but also a park; it is an urban vegetable garden, but also a futuristic machine. Price yields to the metaphor of the design of the park as the design of the city. The park establishes new codes and is proposed as an "operational model of the future", "a test-bed for future voluntary social engineering". The role of the project is to be a precursor, an "anticipatory design", since the slowness of architecture prevents it from being an efficient "problem solver". Its most important function is the achievement of a calculated uncertainty[32]. The park is the concept of the metropolis of the future, and its landscape, almost thirty years later, appears even more advanced to us today.

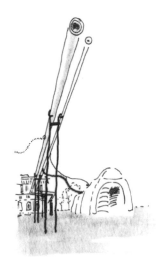

the high-level electronic/pneumatic servicing ribbon . . . .

Procedural diagrams / 2.   The diagrams of Tschumi contain a strong dimension of representation of the complexity and non-sequential character of the contemporary city, but the grid of the red *folies* constitutes a set of points of attachment, orientation, and layout of equipment and services that gives the project legibility. Observing the evolution of Tschumi's project over time – from the first hypothesis inside a different program (1976) through the proposal of 1983 and its successive refinements, all the way to the completed park – we can see the increasing precision of the design that is distilled, structured, and clarified. It is almost as if the initial deconstruction, through the two operations of "disjunction" and "dissociation" of the program, were striving to reassemble in a new order: from a set of independent forms and interrupted routes, without relations, to the superposing of structuring and unifying axes.

ordering and tuning up the familiar.

The project by Rem Koolhaas is an attempt to represent an idea of complex, non-monolithic hierarchy. The complexity is pursued through organization in bands and the coexistence of different groups of urban materials, each of which is integrated in its own system and does not prevail over the others. Each group of elements maintains its own recognizability and autonomy, even on a formal level. The superposition of the groups of materials (the bands with trees and those with equipment, the small furnishing elements, the "confetti", the large existing buildings and those required by the program, the circulation elements), each arranged according to a specific rule of order, does not immediately reveal the structure. The rapid succession of even very different activities creates a multi-faceted spatial quality. This point also reveals the main difference between the three groups of diagrams. The latter, that of Koolhaas, accumulates activities, while the second, of

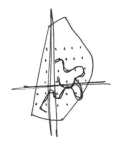

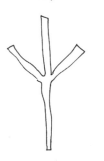

Tschumi, is concerned with generating unforeseen encounters. The first, that of Price, stages the continuous modification of metropolitan space, movement, and its support, at the service of practices.

The impact of the diagrams of Koolhaas, which narrate a tactic more than a project, has been remarkable. Charles Jencks writes that "even unbuilt, the scheme has had an impact on urbanism and superposition theory, because it shows so clearly the consequences of taking different systems and allowing them to run through and on top of each other without trying to synthesize or predetermine the interactions"[33].

The complexity is above all governed inside the succession of bands placed side by side, guaranteeing the presence of different environments. The project's main qualities are the blurring of categories where each part works in the next while maintaining its own integrity, and the complexity of brutal juxtapositions without elements of transition or relation between the parts. Superposition, according to Jencks, is something that goes beyond zoning, the *tabula rasa* and monoculture, and he wonders: is this complexity or complication ? Deconstruction implies something that goes beyond mere fragmentation; fragmentation can carry with it nostalgia for a lost past, but it can also lead to the reduction of a complex system into a simpler system.

Tschumi and above all Koolhaas are closely linked to the reflections of the artistic avant-gardes of the early 20th century, and their path of research proceeds in continuity with the main moves and thoughts that extend from the elementarism of the 1920s all the way to contemporary deconstruction (VIGANÒ 1999). Price pursues a different design – one of social emancipation that also happens through spatial and technological devices. His project seems distant from the contemporary character Tschumi and Koolhaas try to represent, for which they aim to grasp and form the aesthetic: further back, inside an idea of strong technology that is passé, "sixties". At the same time, however, Price's project-concept is catapulted forward into an ideal and distant space in which new dimensions can be explored.

It is not important here to decide if the reflection on complexity conducted by Tschumi, Koolhaas, and Price is innovative and original in the absolute sense. What interests me is to observe how an existing disciplinary background, a concrete site, a series of requirements demanded by society and collected in competition guidelines react to the concepts proposed, and vice versa. The diagrams illustrate some of the possible forms of this relationship.

In a famous sketch, Le Corbusier juxtaposed the image of a TransAtlantic liner with the *Unité d'habitation*, and in so doing he implemented what Alan Colquhoun has defined as "displacement of concepts". In the case of the ocean liner, the idea was to absorb elements extraneous to the high tradition of architecture, "raw elements of the real world". In other sketches, the displacement has to do with the transformation of themes belonging to *high* architecture, such as the traditional three-part division of the building (base, *piano nobile*, attic), and their radical reinterpretation, as in the case of *pilotis* and the roof garden (Colquhoun 1989: 55). The TransAtlantic becomes the paradigm of the new architecture.

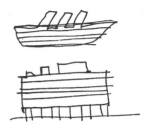

Some time later – and picking up where Colquhoun left off, again with reference to Le Corbusier –, Philippe Boudon distinguishes between *modèle de forme*, for example the *fenêtre en longueur*, and *modèle d'opération* or *modèle opératoire*, for example the *plan libre* (Boudon 1975; Girard 1986). The *patte-d'oie* of the Baroque city represents what Boudon defines as a model, a concept of form. If the term *model* drags with it the idea of the *copy*, I would like to bring out the constructive aspect that permits organization of reality, giving it a form. This form is the order of the discourse that unfolds; it is the sequence of movement, in this case of one or more bodies inside a street that together with others determines the "goose step". The concept has a specific form, but the capacity to define a space of abstraction in which different thoughts and references arise pertains to it. The parks of Lenné, Le Blond, Alphand, along with the concepts of Laugier urged treating the city as a forest with paths and points of intersection.

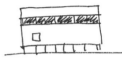

The *modèle d'opération* (again in this case I prefer to translate it as *concept of operation*) is the assembly of an operative structure that investigates reality. "Art does not make what it sees, but as it sees its being made", wrote Quatremère de Quincy in the entry "Architecture" of the *Dictionary*. The concept of operation examines reality and transposes its own modes of organizing itself. In speaking through the project of the city contemporary to each of them, Price, Tschumi, Koolhaas and others are not attempting to imitate what they see, but to retrace some of its modes of construction (Koolhaas 1993b).

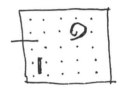

The analytical construction of the project aims at taking over devices – a sequence of operations. Only in certain cases is the operative structure then fully represented in an urban space. In other cases, it remains a diagrammatic form that structures and guides the construction of knowledge by design.

INTERPRETATIVE SKETCHES. CONCEPTUAL SHIFTS: THE *PAQUEBOT* OF LE CORBUSIER, CONCEPT OF FORM, THE *FENÊTRE EN LONGUEUR* CONCEPT OF OPERATION, THE PLAN LIBRE.

[1] See, for example: Forman and Godron 1986; Forman 1995.
[2] Zaitzevsky 1982; see in particular chap. 3, "The Boston Park Movement".
[3] Cited by Zaitzevsky 1982: 35.
[4] Crocker Uriel, H., *Map and Description of Proposed Metropolitan Park for Boston*, 1870, cited in Zaitzevsky 1982: notes 12 and 13, chap. 3. The plan, according to Zaitzevsky, was probably drawn by Francis L., Lee, a Bostonian "landscape gardener".

⁵ OLMSTED 1970: 22. "Greensward" is also the motto of the competition project of Olmsted and Vaux for Central Park.

⁶ Olmsted, F. L., "Seventh Annual Report of the Board of Commissioners of the Department of Parks for the City of Boston for the year 1881", City Document n° 16, 1882, pp. 24-28, published in the anthology of texts by Olmsted edited by SUTTON 1971 (from the paperback edition of 1979: 221-227).

⁷ Olmsted also had extensive influence in Europe. See, for example, FORESTIER 1908.

⁸ Bellamy is an activist of the American Nationalist Party, which promotes socialism based on state-run industry and national solidarity. Also see the third part of this book.

⁹ NOLEN 1916: 165; in the bibliography he includes the text by W. Hegemann on *Amerikanische Parkanlagen*, the catalogue of an exhibition (*Wanderausstellung*) held in Berlin in 1911.

¹⁰ On the opposition to the plan for green areas and the expropriation costs see SCARPA 1983.

¹¹ Patrick Geddes (from a report on "The Town in the Madras Presidency", 1915, Tanjore, pp. 17) cited by SMITH MORRIS 1965: 31. Also see FERRARO 1998.

¹² "As Geddes reminds us, the greenway system depends on an intimate knowledge of the individual characteristics of an area" (SMITH MORRIS 1965: 32).

¹³ See the criticisms of the *City Beautiful* movement in SMITH MORRIS 1965: 27.

¹⁴ LOBSINGER 2000.

¹⁵ ALEXANDER 1965: "A collection of sets form a semilattice if and only if, when two overlapping sets belong to the collection, the set of elements common to both also belongs to the collection. […] A collection of sets forms a tree if and only if, for any two sets that belong to the collection, either one is wholly contained in the other, or else they are wholly disjoint".

¹⁶ ALEXANDER 1965.

¹⁷ Woods, S., "Web", in *Carré Bleu*, n° 3, 1962.

¹⁸ "When we predetermine points of maximum intensity – centres – it means that we are freezing a present or projected state of activity and relationships. We perpetuate an environment where some things are central and others are not, without, however, any competence for determining which things belong to which category. The future is compromised" (CANDILIS, JOSIC, WOODS 1965: 188). This quotation asserts with great clarity the sense of research on the concept of isotropy and the design of isotropy.

¹⁹ GUTKIND 1953.

²⁰ LEFEBVRE 1968. In *Le droit à la ville*, the author emphasizes the ambiguity of the metaphor of the tissue, a "sorte de prolifération biologique" (pp. 11) that can be described by using the concept of the ecosystem, which does not refer only to a form, but "Il est le support d'une 'façon de vivre' plus ou moins intense ou dégradée: la société urbaine" (pp. 12). This way of living brings with it a system of objects and a system of values (among urban systems of objects: water, electricity, and gas).

²¹ LAUGIER 1759.

²² From the report of the EPPV, in BALJON 1992: 37.

²³ BALJON 1992: 38.

²⁴ "The park is a point of encounter of cultures that have the right to individually express themselves: it is a park of reconciliation".

²⁵ Cedric Price, Concours International Parc de la Villette – Report, 1982, CCA Archives, Cedric Price: 156 Parc 1.3.3 Report concours parc de la Villette.

²⁶ For example, in MATHEWS 2007. The obituary published on the day after his death in 2003 recalls his distaste for the common view that saw him as idealistic and not very interested in actually building his projects. There is mention instead of a disagreement on fees that leads to the failure of the possibility of a project for Parc de la Villette prepared in 1986-1989 (http://www.independent.co.uk/news/obituaries/cedric-price-548585.html). A short note on Parc de la Villette appears in Price 2003: the project of a greenhouse in the park. See: HARDINGHAM 2003.

²⁷ In a more accurate reading of the text than I can accomplish in this moment, analysis should be made of the many terms taken from the common language and their relationships with the idiolect of urban planners and architects. The Pop vein expressed, for example, in the choice of images that accompany the text also permeates the choice of words.

[28] "Non-Plan would certainly provide such information. But it might do more. Even to talk of a 'general framework' is difficult. Our information about future states of the system is very poor" (BANHAM, BARKER, HALL, PRICE 1969: 442).

[29] The text of reference is obviously *The Affluent Society* by Galbraith, from 1958.

[30] Cedric Price, Concours International Parc de la Villette – Report, 1982 CCA Archives, Cedric Price: 156 Parc 1.3.3 Report concours parc de la Villette.

[31] Cedric Price, Concours International Parc de la Villette – Report, 1982 CCA Archives, Cedric Price: 156 Parc 1.3.3 Report concours parc de la Villette.

[32] Price, C., "Anticipating the future", *RIBA Journal*, sept. 1981, now in PRICE 2003.

[33] JENCKS 1995, from the edition of 199?: 79.

The construction of a territorial project suggests certain important conceptual shifts for a territory like Salento – a territory that has always been considered marginal, yet is a place of innovative economic forms; not rich in outstanding landscape and environmental features, yet possessing in every part a landscape of great value; that was always considered underdeveloped in terms of infrastructure, but is dense with minute systems that permit it to be inhabited in every part; a territory often thought of as lacking water, which nevertheless floats like a raft on a sea of fresh water that could also supply larger regions; a territory that combines a strong urban tradition with the custom of living in the countryside for certain periods of the year, without making a break between these two forms of dwelling and of organization of social life.

Some of the conceptual shifts impact more general areas of reflection on the territory, but find in Salento exceptional conditions of proof or disproof; others spring from the specific conditions of Salento as an extremity, a peninsula of the peninsula, bordered in its diversity also from the rest of Apulia. Exceptional characteristics, extreme conditions, and diversity make this territory a proving ground for a new hypothesis of territorial design. In order to approach the contemporary situation, this project makes critical use of the tools, categories, and objectives of the design of industrial modernity and, where appropriate, intentionally puts them aside.

VIGANÒ, P., 2001, *SPOSTAMENTI CONCETTUALI*, IN VIGANÒ, P., ED., *TERRITORI DI UNA NUOVA MODERNITÀ/TERRITORIES OF A NEW MODERNITY*, NAPOLI: ELECTA
THE TERRITORIAL PLAN OF THE PROVINCE OF LECCE WAS PREPARED BY A LARGE GROUP COMPOSED OF: P. VIGANÒ (DESIGNER), B. SECCHI (SCIENTIFIC CONSULTANT), S. MININANNI (COORDINATOR *STUDIOLECCEPTCP*); S. ALONZI, L. CAPURSO, A. F. GAGLIARDI, A. D'ANGELO, L. FABIAN, R. IMPERATO, F. PISANÒ, M. D'AMBROS, R. MIGLIETTA): *STUDIOLECCEPTCP*; C. BIANCHETTI WITH P. DE STEFANO, G. PASQUI, L. VETTORETTO (LOCAL DEVELOPMENT POLICIES); M. MININNI WITH S. CARBONARA, P. MAIROTA, N. MARTINELLI, G. CARLONE, G. MARZANO, L. SCARPINA, P. MEDAGLI, L. ROSITANI, M. LAMACCHIA, D. SALLUSTRO (ENVIRONMENTAL AND LANDSCAPE ASPECTS); A. TOMEI (GEOLOGICAL AND HYDROGEOLOGICAL ASPECTS); A. DE GIORGI (ALTERNATIVE ENERGY POLICIES).

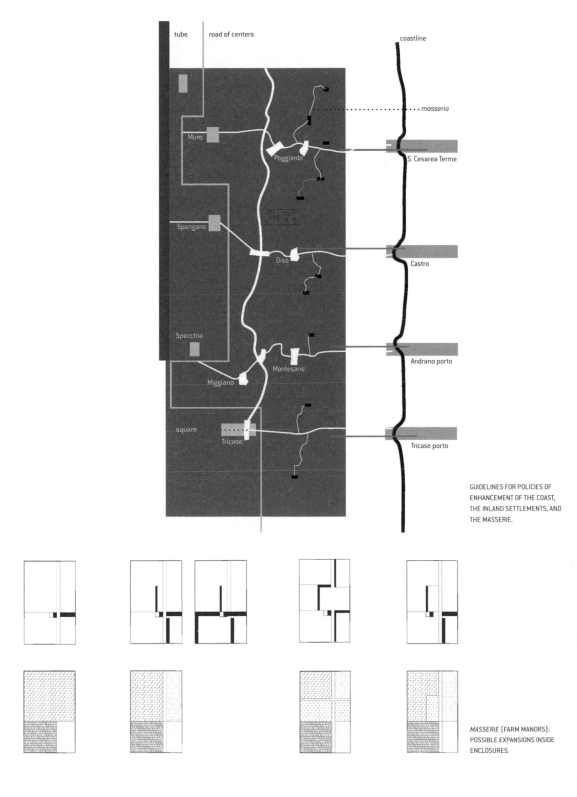

tube   road of centers

coastline

masseria

Muro

Poggiardo

S. Cesarea Terme

Spongano

Diso

Castro

Specchia

Montesano

Andrano porto

Miggiano

square

Tricase

Tricase porto

GUIDELINES FOR POLICIES OF
ENHANCEMENT OF THE COAST,
THE INLAND SETTLEMENTS, AND
THE MASSERIE.

MASSERIE (FARM MANORS):
POSSIBLE EXPANSIONS INSIDE
ENCLOSURES.

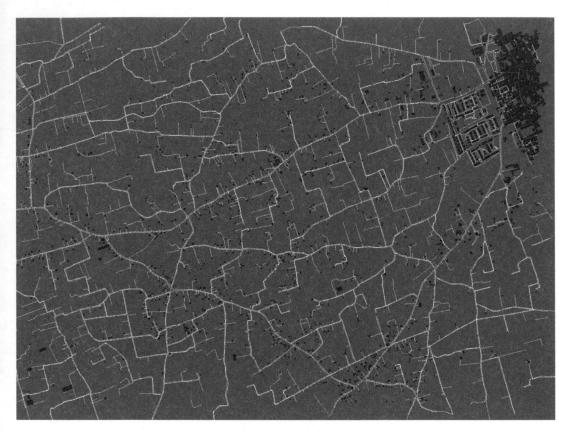

FORMS OF SETTLEMENT
DISPERSION: A PARK-TERRITORY.

Salento as park

The fundamental idea of the Territorial Coordination Plan considers Salento as a large park — the construction of a detailed and complex habitat that takes on the form of a park.

The term *park* is used here in the contemporary sense, not just to indicate a place of *loisir*, but also to point to a set of situations in which environmental characteristics, in the wider sense, contribute in an essential way to construct and enable some or all of the main social practices and activities.

This understanding of *park* implies a slight conceptual shift, not so much with respect to the use of the term in everyday language, as with respect to the importance assumed in recent times by the non-constructed and "natural" part of the environment. While in the 1960s the conceptualization of phenomena of territorial transformation put the accent on the new urban dimension, the rise of environmental questions, an increasingly broad conception of free time, and the desire for living conditions in the midst of greenery have now reversed the perspective.

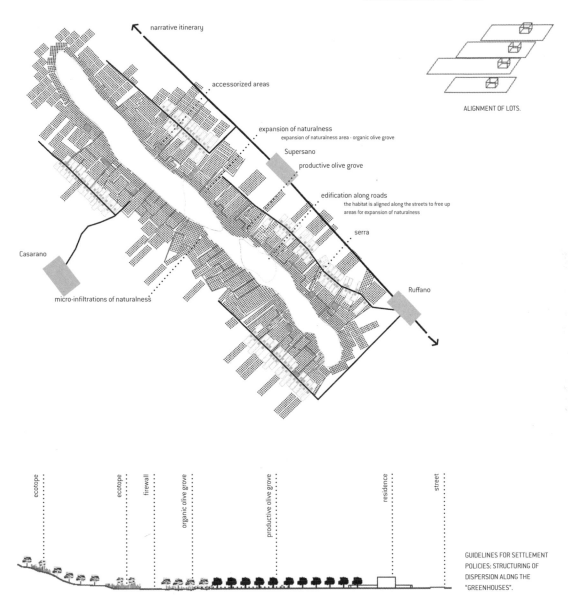

narrative itinerary

accessorized areas

expansion of naturalness
expansion of naturalness area - organic olive grove

Supersano

productive olive grove

edification along roads
the habitat is aligned along the streets to free up
areas for expansion of naturalness

serra

Casarano

Ruffano

micro-infiltrations of naturalness

ALIGNMENT OF LOTS.

ecotope

ecotope

firewall

organic olive grove

productive olive grove

residence

street

GUIDELINES FOR SETTLEMENT
POLICIES: STRUCTURING OF
DISPERSION ALONG THE
"GREENHOUSES".

In the big park of Salento, fragments and segments of a natural context that was once much more widespread coexist, together with smaller or larger urban centers, forming a single variegated, dispersed city. So the idea of the park does not arise in Salento only from observation of its environmental characteristics; it is an overall idea of landscape and the practices that impact it, a pretext to think about the fundamental characteristics of the contemporary city.

Salento is a land of great beauty, blessed like few other zones with a diffused fabric of environmental and cultural assets. It is not a place of concentrated and exceptional landmarks: in the park of Salento several natural and artificial "rooms" of great value exist against the backdrop of a myriad of points of interest, elements of material culture, extended landscapes. This particular condition urges reflection on the potential intersection of urban and territorial policies with those regarding environmental and cultural assets: a park not of constraints, but of potentialities.

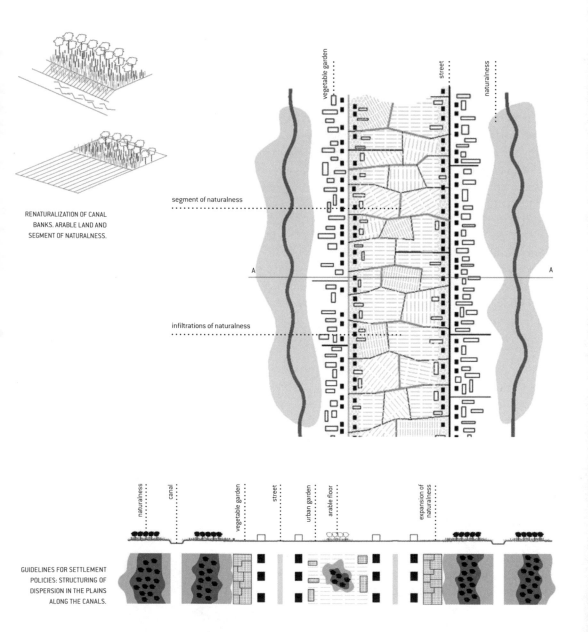

RENATURALIZATION OF CANAL
BANKS. ARABLE LAND AND
SEGMENT OF NATURALNESS.

GUIDELINES FOR SETTLEMENT
POLICIES: STRUCTURING OF
DISPERSION IN THE PLAINS
ALONG THE CANALS.

Diffused naturalness

In the Salento territory it is possible to imagine that the two terms *concentration* and *dispersion* are not opposites, and are both a coherent part of an innovative environmental project.

This project sets out to get away from the two viewpoints utilized in the past. The first asserts that an environmental policy has to focus on a few protected areas; the second, which is more recent, wants to channel naturalness and its spread into precise forms, especially the linear, oblong forms of environmental corridors. The first viewpoint conflicts with the idea of diffusion and reinforcement of naturalness, since the protected areas are just an incomplete sample of territorial resources of biodiversity; the second arbitrarily narrows the field of experimentation of the spread of naturalness and renders banal the concept of the ecological corridor, which, inhabited and utilized by different species, has to be conceived to suit those that will pass through it.

PERMEABLE SURFACES: PATTERN OF TAMARISKS.

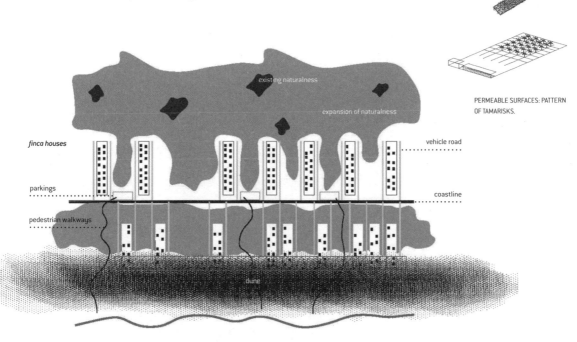

GUIDELINES FOR SETTLEMENT POLICIES: STRATEGIES OF CONCENTRATION OF THE FINCA HOUSES AT TORRE LAPILLO.

The project of the distribution of naturalness includes territories of the land and the sea; it utilizes the few existing fragments of naturalness and proposes their expansion, defining the conditions in which new elements of nature can be born; it investigates concepts of porosity and permeability, expansion, infiltration, and percolation. A project that takes on different forms over time and sets out to mitigate some of the effects connected with climate change, increasing the presence of wooded areas. This landscape project is composed of scattered and random elements.

Environmental cycles

The focus of environmental policy is not only questions of landscape but also infrastructural issues. Providing the infrastructure for a large park like Salento requires exploration of different techniques from the past, as well as innovative ones. The water cycle must be revised entirely and, in particular, has to take into account the problematic process of infiltration of salt water caused by the pumping of thousands of wells. A new model of well distribution that focuses on the

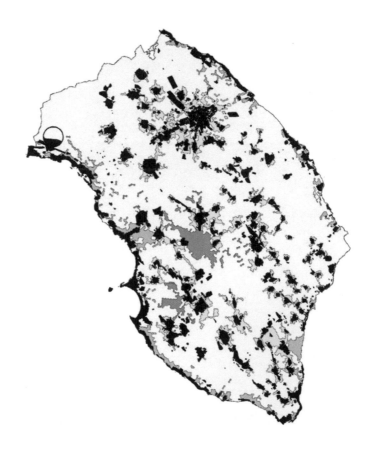

NATURAL WATER PURIFICATION: A SCENARIO: IN BLACK, URBAN CLUSTERS SERVED BY THE SEWER SYSTEM; IN GRAY, AREA OF EXTENSION OF THE SEWER SYSTEM; IN YELLOW AND GREEN, RANGE OF SETTLEMENT DISPERSION TO UPGRADE WITH SINGLE AND CONSORTIAL PHYTOPURIFICATION SYSTEMS; IN BROWN, SUBCOASTAL AREA TO UPGRADE WITH PHYTOPURIFICATION SYSTEMS WITH GREATER LOAD FLEXIBILITY (OXIDATION PONDS AND LAGOONING).

inland strip instead of the coast, as well as the use of techniques of phytopurification downstream from the purifiers and wherever it is possible to recoup
and reutilize purified water. It defines new modes not only of exploitation of
resources, but also of organization of the territory, its modes of concentration
and dispersion.

Loose-weave infrastructures, tubes, and sponges

In light of the lack of certain major infrastructures that would put the Salento territory into communication with the rest of the world, it is under-infrastructured.
This is reflected in greater costs of interconnection with vast and distant markets.
However, Salento is densely infrastructured in terms of its material resources connected with agricultural use, the extension of the rail network, and the reclamations, and this leads to the general inhabitability of the territory.
Cities in Salento are served by a dense network of radial roads directed towards
their center. In past decades the radioconcentric model has been completed
through the construction of ring roads and expressways that have superimposed

THE RAILROAD. THE TUBE + THE
PENDULUM + THE RAILROAD
(TO THE LEFT).

THE SPONGE: THE NETWORK OF
STREET PERCOLATION
(TO THE RIGHT).

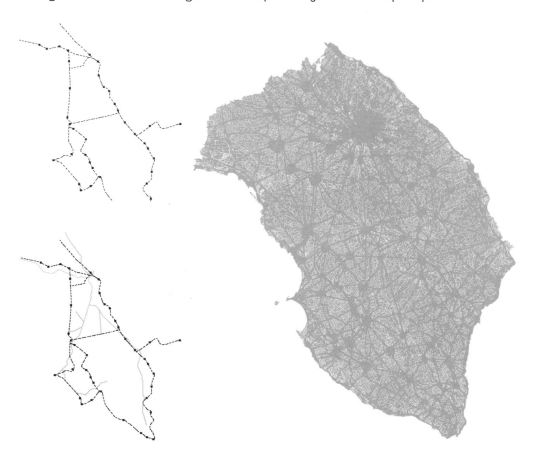

the ring on the radial functioning. The ring presumes that there will be balanced growth of the city in all directions — something that rarely happens; for this reason, the many ring roads in Salento often intersect varying situations that make their impact less definitive than expected.

The *Territorial Coordination Plan* proposes the passage from a radiocentric and concentric urban model to one that is open, that intersects the centers and connects with a loose-weave network that crosses the Salento peninsula. This network relies on a minute fabric of streets, a sort of "sponge" and a few "tubes" devoted to faster traffic. Maintaining different speeds of crossing Salento also means passing from a merely functional idea of the road system as the fastest possible connection between two points, to a more complex hypothesis where the street becomes a place devoted to various practices. The conceptual shift proposed leads to this use and design of a dense road network in Salento as a set of "narrative itineraries" — as one of the most interesting ways to narrate a territory, not just its past history, but also the landscapes that describe and represent it today.

THE LAYERS AND ROOMS OF THE
PARK (TO THE LEFT).

THE ROAD OF THE CENTERS +
THE RAILROAD. THE PARKWAY +
THE NARRATIVE ITINERARIES +
THE RAILROAD (TO THE RIGHT).

While rethinking the water cycle imposes an important conceptual shift by introducing new differences between the various parts of the territory and therefore a new relationships between them, the theme of refuse must move away from a model organized through dumps in a territory such as Salento, marked as it is by innumerable phenomena of karstification. This produces pollution that is hard to control, and for this reason it is necessary to carefully consider alternative solutions that focus on the construction of a network for differentiated disposal, making it possible to correctly size the technologies and facilities for the elimination of waste.

A new development model

Salento belongs to the south of Italy, a part of the country that until now has performed less strongly from the viewpoint of economic growth; but in recent years its economy has shown remarkable signs of vitality in many sectors, and this makes it possible to think about a possible model of growth that does not repeat the steps and itineraries followed in the past by other regions that are now more developed. Innovation is often more interesting than imitation.

If we think of the economic system — also in its territorial representations — as an open system, inter-related to the physical environment and part of the culture of a place, a territorial project is an opportunity to propose certain directions of growth whose reason for being lies in the environmental and cultural characteristics of Salento.

GUIDELINES FOR ENHANCEMENT POLICIES: CIRCUIT OF *NEGROAMARO*.

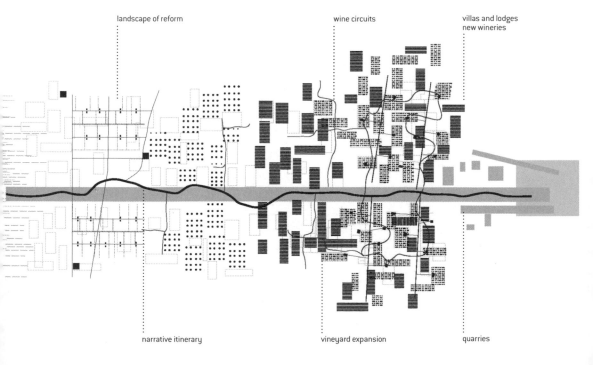

landscape of reform   wine circuits   villas and lodges new wineries

narrative itinerary   vineyard expansion   quarries

The first proposed direction imposes a strong conceptual shift, with respect to the tradition. It imagines a diffused model that no longer calls for great concentration of financial and human resources in a few physical, sectorial, social, or institutional places, and it explores the policies of rebalancing that would result. This model, especially in societies and economies in which small and family businesses are the protagonists, can be equally – if not more – efficient and productive than a model based on "poles of development". The project sets out to consider the formation of the diffuse city in Salento, which is described for the first time as a settlement form completely consistent with this model of growth. If correctly designed and governed, the form is aligned with a correct environmental policy and the spread of naturalness, and also with the safeguarding and conservation of the important cultural role of the urban structure of Salento.

The second direction, which pays attention to fixed social capital in the wider sense of the term, also requires slight conceptual shifts. It proposes, for example, setting up a policy for tourism as exploration of the differences contained in the Salento territory, thereby extending the depth of coastal tourism into the inland zones. In definitive terms, this means taking advantage of the many centers, the many *masserie*, i.e. of the dispersed model, and considering diffusion as an opportunity to achieve higher living standards and more versatile territorial distribution of equipment and facilities.

The third direction should not actually be limited only to the region of Salento, but should necessarily come to grips with national policies as well. This consists in imagining that the climate conditions of Salento, like those of much of the Mediterranean, make it possible to immediately envision an energy policy that is radically different from that of the past. Therefore the project outlines several energy scenarios intended to open a discussion on concrete possibilities for renewable resources and on the consequences of the adoption of a different energy model on territorial organization: from a network model to a decentralized one, particularly suited to diffused growth and an equally diffuse city.

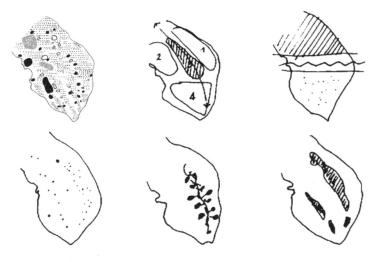

CONCEPT: ROCKS AND SPONGES (ON TOP).

THE PARTS: THE RIDGE.

SOCIAL LANDSCAPES: NORTH\ SOUTH.

THE LANDSCAPE OF SENIOR CITIZENS: HISTORICAL CENTERS THE HABITAT OF ACCESSIBILITY CONURBATIONS (BOTTOM).

A NEW ENERGY MODEL.

There is a moment in which the project, prior to being defined in all its parts, is conceptualized and provided with a diagrammatic, schematic, abstract, and meta-phorical representation that is not always measurable or to scale, often pertaining to relations, and is therefore topological. A representation exists of the projection set in motion by the project, which indicates the degree of detachment from the contingent. The above examples, even if they are often simplified and reduced, have a long tradition in the Occidental design of the city and the territory. The con-cept is not the schema of the design, or the design of an overall sale, but the con-tribution the project is able to offer to a rethematizing of a given problem. It is both the thesis and the hypothesis. It is the shift in viewpoint, the *décalage*, the *derive*. Alongside a return to experience, to subjective knowledge that has inspired much research on contemporary space, there is also a return to modernity as a process of objectification (BENJAMIN 1936).

Deleuze and Guattari reveal the multiplicity of the concept: "Every concept has an irregular contour defined by the sum of its components, which is why, from Plato to Bergson, we find the idea of the concept being a matter of articulation, of cut-ting and cross-cutting. The concept is a whole because it totalizes its components, but it is a fragmentary whole" (DELEUZE, GUATTARI 1991). The concept approaches the magma of reality and traces back to problems "poorly considered" or "poorly formulated". Every concept "has a history" which is often non-linear, zigzagging: in a concept "there are usually bits or components that come from other concepts, which correspond to other problems and presuppose other planes. This is inevita-ble because each concept carries out a new cutting-out, takes on new contours, and must be reactivated or recut" (DELEUZE, GUATTARI 1996: 8). It should not be con-fused with that in which it comes to be. "It is an act of thought" (DELEUZE, GUATTARI 1996: 12).

The diagram is the "abstract machine", mute and blind, as Deleuze writes regard-ing the term used by Foucault in *Surveiller et punir*, though it is what "produces sight" and "produces speech" (DELEUZE 1986: 42). Every society has its own dia-gram or diagrams – modes of functioning and materials not yet formalized in which relationships of force and strategy are expressed. The concepts of urban-ists and their diagrams do not fall outside this interpretation.

Being more than the result of a process of interaction or problem solving, the con-ception is the outcome of operations. "Possible objet de connaissance" (BOUDON

2004:22), conception can also bring together different projects, behind which similar operations are located. Conception as a cognitive process underpins the positions of Design Research where the generative process takes on greater centrality than that of the hypothetically concluded and conclusive form. If the process, in its various interpretations, asserts itself as an individual history, a singularity, the idea of operation has a more pronounced generalizing content, capable of holding together even very different objects. It forms a communicable and shared layer that detaches from the idea of creation as incommunicable and links to an individual poetics, though without negating it.

"What happens between the moment in which a resistance becomes perceptible, in which the obstacle is revealed, and the moment in which the problem is formulated? A space of representation is glimpsed in which the terms of the problem can be inscribed" (ANDLER 1987). The need arises to inscribe these objects within a space of abstraction, a fictional space in which the operations of its resolution can be performed. Here, I have discussed this space and its contribution to the construction of knowledge.

TERRITORIES OF DESCRIPTION

Part II

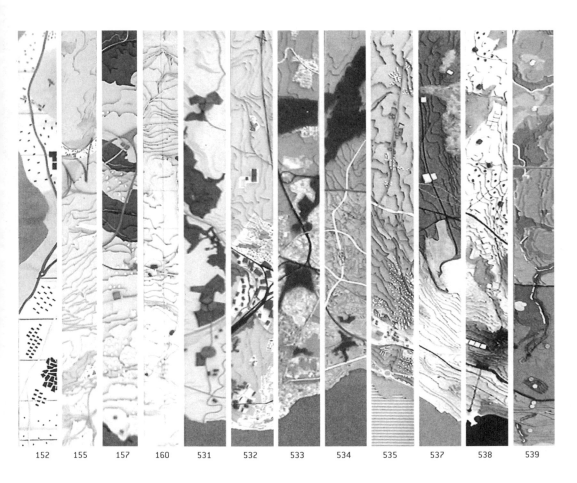

152   155   157   160   531   532   533   534   535   537   538   539

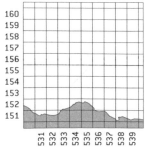

X-LAND SUISSE
EPFL DESIGN ATELIER (P. VIGANÒ
WITH M.P. MAYOR, C. BEUSCH)
LAUSANNE 2003–2004.
THE TERRITORY OF LAUSANNE
IS DIVIDED INTO SECTIONS OF
1X11 KM, EACH STUDIED
INDEPENDENTLY. THE MODELS
ARE ON A SCALE OF 1:10.000.

160
159
158
157
156
155
154
153
152
151

531 532 533 534 535 536 537 538 539

The second section of this book focuses on design as a producer of knowledge through operations of description. It concentrates on the ambiguous space that exists between description and project. In other words, the theme examines the unavoidable dimension of interpretation and construction of the description, as well as the descriptive dimension of the project, its equally unavoidable *capacity to describe*. What interests me about the descriptive dimension of the project is its relationship to the characteristics of the space, the selection it makes, the representation it provides. My hypothesis is that the project should be taken as one of the cognitive strategies of contemporary territories and societies; its descriptive capacity is of crucial importance to legitimize it as a place in which the interpretation of the city and its changes is constructed.

This second part is divided into three chapters. The first, *New Territories: A Meta-description*, investigates the terms over the last few decades for the construction of a project space and a discourse on the contemporary city starting with recognition of a new theme, the territories of dispersion, and from the selection of places and descriptive operations. The second chapter, *Design As Description*, more directly approaches the theme of the descriptive capacity of the project. It is a tool of selection, discoverer of landscapes, surveyor of long-term structures, often translating into images that are simultaneously descriptive and normative. The third chapter, *Projects That Describe*, continues to delve deeper into the hypothesis of the project as producer of new knowledge through operations of description. In two investigated territories (*Territory 3, The Porous City; Territory 4, Elementary*) some of the issues raised are explored further.

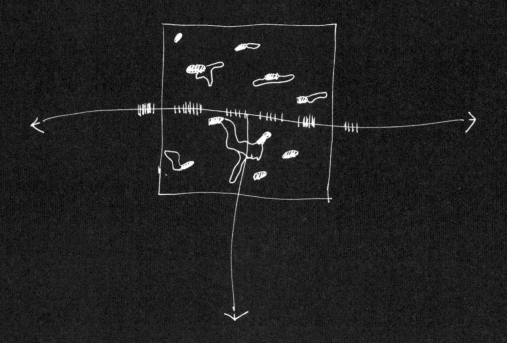

campagna urbanizzata

I'd like to start with a type of description that has been extensively studied: literary description – rhetorical as opposed to narrative form. According to Jean-Michel Adam who has written extensively on description (ADAM 1993), there are three main reasons behind the critique and often the rejection of description: description as imperfect definition, description as impossible copy of the world, description as something disorderly and arbitrary. I believe these elements of critique should be overturned today and examined with renewed interest.

First critique. For the logicians and grammaticians of Port Royal, description is an imperfect definition. While the description allows us to recognize an object as opposed to another one, conveying an idea of its characteristics, it does not define the object in depth because it does not demonstrate its essential attributes (in this sense description is not definition). According to Abbé Mallet, who contributes the entry for the *Encyclopédie*, it does not respond to the question *Quid est?* (What is it?), but to *Quis est?* (Who is it?). Description permits us to recognize not the family, the genus of belonging, but the individual itself. Perhaps the present proliferation of descriptions, then, has to do with the process of individualization. Together with the sensible, the inexact, the individual, the description represents values considered negative in the theological or rationalist perspective of the French classical age (16th and 17th centuries), which instead are now emerging in modern and contemporary society.

Second critique. Description as an impossible copy of the world: Adam cites a story by Yourcenar, *Comment Wang-Fô fut sauvé*, contained in the *Nouvelles orientales*[1]. The story stages the paradox of the perfect description. The paintings of the artist Wang-Fô deceive the young emperor isolated from the world and disgusted by the encounter with things themselves, which are much less tolerable than their representation on canvas. The painter runs the risk of paying with his life for the gap between description and reality, for his capacity to see, read, and represent nuances. The awareness of the impossibility of an exhaustive, thorough description – since it is always the expression of a partial viewpoint, though perhaps one that is sharpened and capable of going in depth – leads to the need for plural descriptions, multiple viewpoints.

THE URBANIZED COUNTRYSIDE: INTERPRETATIVE SKETCH OF THE IMAGE, PROPOSED BY G. SAMONÀ.

Third critique. If narration is a composition that forms a whole whose parts are interconnected by plot, description eludes these rules, seeming always arbitrary, disorderly, and capsizable, risking useless detail. These weaknesses of the descriptive apparatus are remedied, as has been done over time, by specifying schemes of description, guided routes of restitution of the real, in pursuit of rigor and shared standards. In the literature of the French classical era, the description of a tree began from the roots and the seed; that of a man began from the face, the only place in which it is possible to recognize the individual. These are arbitrary rules that conceal a system of values, that express a culture in which it is the face, the noble and least animal-like part of the human being, that is the primary object of description, coming prior to the rest of the body and the garments that only temporarily distinguish one individual from another.

The three critiques have had different levels of importance over time, and have been marked by successive changes in Occidental culture. An attempt to explain the proliferation of descriptions, in particular description of the present time and the present territory, which abound so prolifically over the last twenty years, can find a starting point in the reversal of the criticisms applied to the descriptive operation — overturning what has been indicated as an expression of weakness into an element of force and necessity.

An initial, strong and un-deferrable necessity has been to exit the grouped frames in which Western modernity has attempted, often insistently, to insert multiple individualities, not only of persons but also of places, objects, sensations, ways of life. Description, freed of the pressures of narrative, has enabled many to gain knowledge of what was being transformed around them; widespread operations of sampling[2] have constructed a provisional "Chinese encyclopedia" that in an episodic, fragmentary manner — without continuity — has often revealed unexpected novelties or simply little-known and neglected aspects.

The second necessity, linked to the first, has been that of multiplying viewpoints: if the description is an impossible copy of the world because it not only fails to define the world but also inserts each individuality, revealing others, then only a multiplicity of descriptions can move towards a richer panorama, if not an exhaustive description, closer to reality and to the many individuals that are its components. Many descriptions attempt to convey the unfolding of an experience of the city, juxtaposing narrations that identify places. The multiplication of descriptions is, to a certain extent, also multiplication of theories regarding the real (MALFROY 2000)[3]. Thus the abundance of descriptions bears witness to a wider-ranging effort at comprehension and theoretical construction that impacts the contemporary world.

The third critique — description as construction of arbitrary schemes, or the risk of repetition of clichés — can be reversed via the current necessity for "descriptions of descriptions"[4] that observe their progress as a cultural phenomenon indebted to wider values and ideologies that differ across time and in different places. The description of clouds is no longer the one contained in 19th century serial novels, and while it is true that many descriptions tend to flatten the individuality of each phenomenon in unoriginal reports, stiffening the form of clouds into the usual "majestic sailboats" or the periphery into "foam gathering on the shore of the city"[5], in a formless, chaotic mass without qualities, it is also true that a meta-description can reveal something not only about unobserved practices and places, but also about the movements of our culture. After years of attempts to describe the contemporary city, it is urgent to begin describing the character of what has been described, to observe what has been included and what has been left out, to recognize the new clichés that have replaced the old ones, and their reciprocal relations and distances.

There has been much discussion of description in recent years[6], as the expression of a return to experience as the primary source of knowledge, not preordained, starting from the difficulty of making images, culture, and disciplinary knowledge jive with what we encounter out there when crossing a territory by car, on foot, or by train. Many sense the embarrassment of the gap between observation that attempts to use the available tools, the established categories, such as "city" and "country", for example, to describe an object, and the vast territories of settlement dispersion which feel so alienating for many scholars. Architecture and urban planning seem to no longer be living in the city and the territory; we meet them only by chance, reduced to fragments of ideas and projects or, more often, in the version that has seemed to many to represent their failure. The idea of the impossibility of the practice of description, of the indescribability of the city and the futility of attempting to interpret contemporary urban phenomena, the retreat into an utterly internal form of knowledge, has emerged in the debate of recent decades alongside efforts and attempts that move in the opposite direction. Many of the projects of recent years have taken these positions as a way of legitimizing their own choices – project-fragments, introverted or nostalgic projects that try to oppose ideas of the city and architecture against something that does not appear to be recognizable, but certainly cannot be shared. The judgment has frequently been expressed in terms of prejudice regarding an object, the contemporary city, that has often not been observed or understood. In this context certain places, situations, and operations have played a crucial role in the construction of a discourse on the contemporary city, in the recognition and invention of new territories. The selection of situations, images, figures, metaphors, descriptions, and stories marks the development of a discourse: a sequence of arguments that structures the interpretation and becomes the medium through which, and in which, the interpretation takes form. The impossibility of a naive gaze is evident, just as the interest of the deconstruction of discourses and the revealing of their underlying ideologies is clear. I cannot, in this moment, deconstruct the discourse[7] on the contemporary city developed over the last twenty years, but I believe that in the near future, this effort will be necessary.

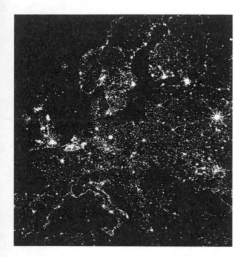

EUROPE BY NIGHT.

Places of description

The territories of settlement dispersion have been at the center of both the debate and the attempts at description/interpretation of the city over the last few decades in Europe. Certain representations, such as the image of Europe by night, have revealed unseen continuities and proximities between places thought to be separate and distinct in the past. The territories of settlement dispersion have played an important role in the formation of a discourse regarding the contemporary city, proposing a new object of research and design. Their description has been the "creation of intellectual territories" (CRYSLER 2003: 195) and the first design move that invents and introduces the object as a relevant theme, together with new scales, subjects, landscapes. Therefore it has been necessary to update the definitions, to produce original surveys, images, and metaphors, to convey a blurred picture that is never fully understood.

Territories of dispersion.   Until the interpretation of the new settlement condition as a large "diffuse city" (INDOVINA 1990; SECCHI 1991), the readings and images of dispersion produced in different contexts were guided by the persistence of the term *city*; only more recently has certain research proposed putting the territory at the center, thereby modifying the viewpoint on the observed phenomena. While the continued use of the term *city* (though accompanied from case to case by an adjective that selects the quality deemed relevant) often implies the existence of dynamics capable of reinforcing the characteristics traditionally attributed to the city, namely density, centrality, urbanness, even when they exist in configurations that differ from the established ones, the "territories of dispersion" abandon the arduous search for an analogy with the urban phenomenon and interpret the new settlement condition as something "other than city". These territories have been observed as places of rupture and dissolution of the city, but also – in keeping with the interpretation of certain Italian economists and geographers between the 1960s and the 1990s – as a phenomenon that has reduced contrasts, making a territory more cohesive, introducing new forms of production and habitat that are scattered, specific, and individual (BAGNASCO 1977; FUÀ and ZACCHIA 1983; BECATTINI 1997; 1998).

The two interpretations – that of the break in a supposed previous balance between city and country, on the one hand, and that of the continuity, the "development without fractures" reinforced by widened utilization of the territory permitted by its extensive infrastructuring, on the other – have pursued each other for at least two centuries in Europe, bringing forth different social strata, economies, ideologies, and policies. The image of isotropy as a form of perfect democracy did not only fascinate Wright in Broadacre City, or Gutkind in the diagrams that accompany *The Twilight of Cities*; it is also present in many European territories, and it is simultaneously physical construction, landscape, expression of an imaginary, and political choice.

Compact city, diffuse city, and territories of dispersion are categories of analysis that also indicate different lifestyles, imaginaries, and social relations; nevertheless, a continuous flux of hybridizing passes from one to the next in an overall redefinition of contemporary ways of living together (BARTHES 2002). From this viewpoint, the territories of dispersion appear to be places of formation of specific idiorhythmics that differ from those of the large modern suburbs or the city, though the settlement figures, from the fabrics to the filaments of isolated houses with yards, to the large developments, cross the various conditions. In the dense city, spaces take form for the idiorhythms typical of the territories of dispersion, and the simulacra of urban centrality arise inside them (PELLEGRINI and VIGANÒ 2006). The construction of new inhabitable places in which different rhythms can coexist and generate new spaces of sharing, autonomy, and appropriation is the task on which we must reflect today. The territories of dispersion are one of its main fields of investigation.

Places and possibilities

The descriptions of the territories of dispersion through long efforts of mapping (CORNER 1999) and field research have shed light on the extraordinary possibilities contained in these places and others. It is in the act of description that the hypothesis of possibility in relation to a given *milieu* is structured. The description, asserted the French geographer Vidal de la Blache, aims to clarify the relationships between man and nature in terms of possibility. The focus on the *milieu*, the environment, is not made rigid by deterministic hypotheses, but relies on a *possibilisme*, a concept close to the more recent constructivism or interactionism that places interaction between processes of a different nature at the center of the contemporary sciences (BERDOULAY 1993: 26). An approach that is similar in many ways also surfaces in the surveys of Geddes, from more or less the same period[8]. The project as description interprets places and, more in general, *milieux*, as a field of possibility existing to a great extent prior to the programs in which it is involved.

Milieu.   The definition of *milieu* opens the text *Principe de Géographie humaine*[9], published from the manuscripts of Paul Vidal de la Blache after his death in 1918. Calling on Ratzel, the first geographer to focus on the idea of general facts connected to the globe conceived as a single whole, Vidal traces the facts of human geography back to the relationship with a *milieu* that constitutes the combination of the physical conditions of which the earth's surface is the overall picture. The notion of *milieu* comes from botanical geography that during the course of the 19th century opens up new paths of research that put climate and vegetation, soil, temperature, and humidity into relation with one another. Von Humboldt had already partially glimpsed this approach, and had been the inspiration behind

the *Physikalischer Atlas* of Heinrich Berghaus published in 1836, which was supposed to illustrate his *Kosmos*. Together with Berghaus and Ritter, Von Humboldt had also founded, in 1828, the Berlin Geographical Society (*Gesellschaft für Erdkunde*). The field of ecology opens up through botanical geography, Vidal writes, citing Haeckel who had introduced the term in 1866, defining it as: "the mutual relations of all living organisms in a single and same place, their adaptation to a *milieu* that surrounds them"[10].

Vidal's concept of *milieu* or *environnement* is fascinating: far from an idea of homogeneity of places and landscapes, the *milieu* is the product of heterogeneous, composite parts. The force of the *milieu* lies precisely in its capacity to hold together "heterogeneous beings in cohabitation and reciprocal correlation"[11], designing the geographies of living things. The artificiality of each situation and the effort of adaptation of disparate beings generate associations of elements, the cohabitation of invaders and survivors that gradually adapt to a shared life. This is why Vidal sees the density of settlements as a relationship of fundamental importance, and high density as an utterly temporary condition generated only by favorable circumstances. The behavior of plants is a metaphor for human behavior: "Just as plants dry up when warmth and dampness are lacking, so in analogous conditions human groups tend to retreat". These are the terms Vidal uses to describe molecular groups, discrete formations, in the chapter *Formation de densité*[12].

The idea of the *milieu* is synthetic: it is a field of action that permits study of the processes of individualization and differentiation, observing visible movement and continuous change. Observation is the first step of the cognitive process. Alongside the concept of *milieu*, that of *genre de vie* constitutes the intermediary between the dense material of latent possibilities and the creation of new *milieux* that accumulate and are perfected over time. They are the contingent results of what has happened previously along a chain not deduced from necessities, but from possibilities. *Milieu*, a notion dear to Taine[13], *environnement*, and ecology are concepts that all make reference to the problem of coexistence of multiple groups inside the same space, and to the interdependence among various animal or vegetable subjects. Vidal recalls it in an era – the years around World War I – in which "L'homme de nos jours n'a d'yeux que pour se contempler dans l'exercice de sa puissance" (VIDAL DE LA BLACHE 1922: 105-106).

Human geography makes use of ecology, initially part of biogeography, shifting geography itself away from its relationship with history and depoliticizing it (ROBIC 1993). It establishes analogies with botanical and zoological geography, extending and assimilating the principles, methods, and concepts of ecology, the science of the *milieu*. The earth sciences have a great influence on Vidal[14], who, in the *Tableau de la géographie de la France*, takes the geological base as a fundamental structural factor in the development of the *milieu*. From the *paysan géologue* (VIDAL DE LA BLACHE 1903), to the relationship between the soil and local life, Vidal develops a language of thresholds and crossings whose basis lies in geological characteristics[15]. It is in geology and multi-scale analysis that the hypothesis of

THE TERRITORIES OF SETTLEMENT DISPERSION: VIEW OF THE PLAIN BETWEEN CASTELFRANCO VENETO AND BASSANO.

France as a territory of differences can be situated. It is an interface between the Mediterranean and northern Europe, a country of transitions, shadings, and borders marked by different landscapes. The rejection of simplified and caricature-like oppositions, such as that between the north and the south of the country, is also rooted in this geological interpretation. The set of the differences does not form a mere aggregate, but "a system of places, for which the notion of situation (or *Weltstellung*, taken from Ritter) is the key notion"[16].

Idiography

With Vidal, description becomes the starting point of a new approach in which, without losing its analytical properties, it becomes a synthetic, selective tool of interpretation, observing movement and dynamics. Rather than from the con-

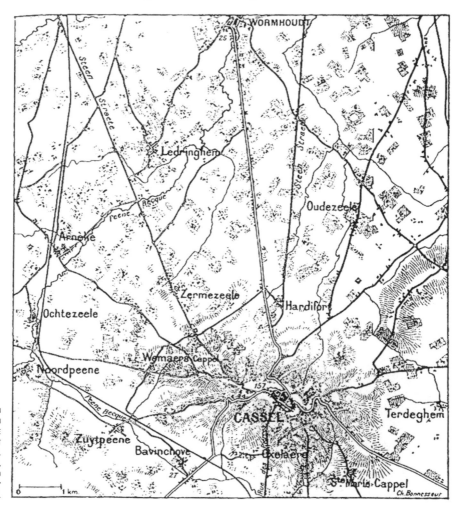

"TYPE OF SETTLEMENT IN FLANDERS, CASSEL REGION (*PAYS DU BOIS*). THE FARMS ARE THE ESSENTIAL SETTLEMENT TYPE HERE (DE PEUPLEMENT). HUNDREDS CAN BE SEEN, FREELY SCATTERED, BEYOND THE STREETS [...]" (VIDAL DE LA BLACHE 1903).

struction of causal chairs or the application of abstract principles, it is from recognition of the individual that it is possible to arrive at shared elements, through
analogies and comparisons. The categories of interpretation are developed during
the course of the description itself, which is a slow, lengthy process of making
contact with the object grasped in all its complexity. Vidal's hypotheses often take
form starting from *indices* that emerge through the ability to read the vegetation,
to understand the role of water, the nature of the soil, the nuances of the climate.
The clue reveals by means of contiguity; the chain of hypotheses is genealogical,
starting with the geological situation tracing back to remote times, or it is relational
and horizontal, seeing the situation as the product of the constitution of individual
and specific relationships. "Each new region is presented as an enigma to be clarified"[17] by the practiced eye that glimpses what generic visual experience cannot
grasp. The clues are precisely the signs that betray the difference: between the

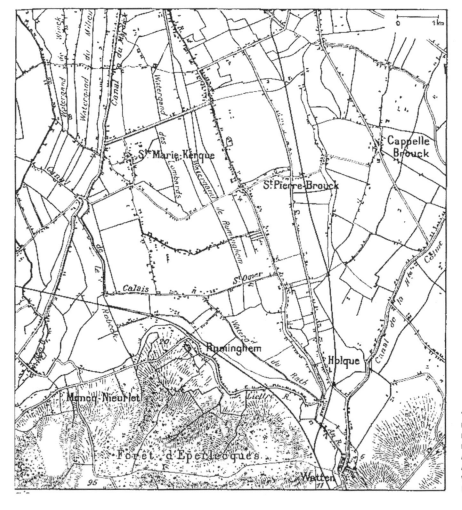

"TYPE OF SETTLEMENT IN
FLANDERS, WATERGANDS
REGION (NOORD LAND) [...]
ROWS OF HOUSES ALONG CANALS
AND STREETS [...]; SOME,
AROUND A CHURCH, FORM A
VILLAGE EMBRYO"
(VIDAL DE LA BLACHE 1903).

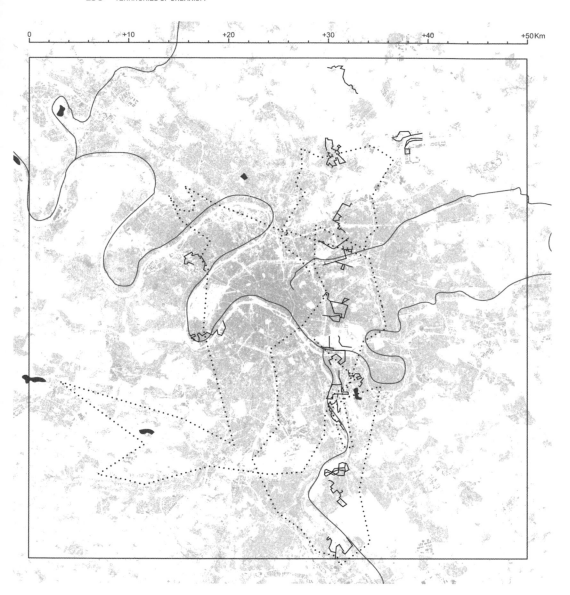

The city is divided in two part. In fact before Ris and Orangis were two cities separated by N7 that was a national street. Now the street is urban and the only link between the two cities.

When the RER D arrived built a separation in between Maison Alfort / Alfortville, from this side are the poor and from there are the rich.

I arrived in September. In apartment there is lack of interaction between us.

This was at the beginning a neighbourhood built for the low middle class. Now there are a lot of immigrants and also the social class is lower.

plains of the North and the Parisian basin, the difference "does not impose itself on the gaze, but betrays itself in numerous signs"[18].

In the *Tableau de la Géographie de la France*, Vidal distinguishes the various settlement forms of the French territory. Territories of concentrated habitation emerge, as do others of dispersed habitation, such as rural Brittany or Flanders, already described by Guicciardini, cited here by Vidal as "une ville continue"[19]. The maps show the relationships in a given period between the infrastructural support and the form of properties, agricultural fields, settlement types and ways of living together outside the urban center. Today, a history of dispersed habitation in Europe – that no critic of the city or the territory has yet felt the urge to write in a complete way – would have the merit of reconnecting more recent phenomena of dispersion and diffusion[20] to many different long-term histories of the European territory, to the interpretations of historians and geographers (or philosophers like Rousseau, regarding Switzerland) who have described them in past centuries, especially from the 1800s to the early 1900s.

Idiography, as a recording of differences, examines the individual and specific qualities avoiding reduction to a version or variation of a type. Therefore, the individual cannot belong to a determinist perspective in which he will, in any case, be an exception. The concept of *genre de vie* is the meeting place of a land and the practices that impact it. Like a plane on which the traces of all the phenomena that have affected it remain visible, the earth's surface gathers the imprints inscribed upon it that modify its plastic characteristics (Corboz 1998).

Operations of description.   The central importance of the operation of detection in the cognitive and rhetorical strategies that have contributed to create the "intellectual territory" of the contemporary city is clear[21]. Approaching the contemporary territory, initially by crossing it, has seemed necessary and inevitable to many people over the last twenty years. Writers, urban anthropologists, photographers, directors, and painters – perhaps before the others – have started to describe their paths in the subway (Augé 1986), from one station to the next of the RER (Maspero 1990), along highways, on the outskirts of the city (Morier 1994), along market-roads (Lanzani 1991), following an intuition, more than a starting premise, that a society composed differently from that of even a recent past must be investigated anew, in ways that are partially different from the previous ones. These ways, starting from multiple viewpoints and techniques, are all closely connected to the inspection, the survey, the walk, whose depth is not just literary. Consider the Surrealists of the 1920s, the Situationists and their *dérives* in the 1950s, or the work of Chombart de Lauwe[22]. In these cases the field survey is necessarily incomplete, conducted by researchers conscious of the veil existing between the gaze and things, but interested in direct contact, convinced of the need to also see with one's own eyes, to listen and to touch (Lévi-Strauss 1993). "To live in a city", writes Julien Gracq in *La forme d'une ville*, "means weaving one's daily peregrinations into a maze of paths usually linked around several directional axes" (Gracq

"PARIS PAS À PAS, LES PROMENADES DANS LE GRAND PARIS" (STUDIO 09, SECCHI-VIGANÒ WITH THE STUDENTS OF THE EUROPEAN MASTER IN URBANISM, EMU).

S. MURATORI, STUDIO PER UNA
OPERANTE STORIA URBANA
DI VENEZIA, 1960. DETAIL OF
THE SURVEY OF THE FABRIC OF
CAMPO DUE POZZI.

1985]. The contemporary practices of wider use of the territory construct new geographies of extended paths that determine a city different from that of the past.

The practice of the expedition, direct observation, or jotting things down in a notebook becomes a strategy of approach to the material character of reality, and in recent years has taken on an unprecedented breadth that bears witness to the widespread degree of discomfort of those who try to understand the present and to imagine the future of the city. To describe means not only to observe but also to represent the observations that have been made. This requires continuously moving between abstraction and experience, images and concrete dimensions, selection and overall representations. The relative duration or inertia of these reflections reinforces my conviction regarding the central importance of the operation of detection and, by extension, of readings of contexts in the cognitive and representative strategies of the project. Idiography, the writing of differences, has contributed in substantial ways to the emerging of the 'contemporary city' and, as I will assert in the third part of the book, a solid bond connects the capacity to read the *possibilities* of a *milieu* and the construction of scenarios for the future of the city.

Surveys, theories

Deformability limits.   In 1959, Saverio Muratori publishes *Studi per una operante storia urbana di Venezia* after a decade of courses[23] in which the "field study" shifts from marginal exercises to become the central, substantial activity of the teaching ("school of the real")[24]. Muratori denies heterogeneity, reacting "to the loss of the notion of connection, hierarchical ranking", to the "arid technicism", the "abstract and sterile aestheticizing ravings", and delves into surveys to reconstruct the evolutionary process of the "urban organism". The operation of surveying, "still so often considered […] a mechanical work of measuring and generic description" (MURATORI 1960: 13), seems to offer a way out of a sloppy and generalizing approach. It is subjected to ceaseless critique and methodological research, and it accounts for the fact that the nature of the phenomenon being studied is always deformed by the device of observation and measurement, by the technical intervention of the investigation (extending, to a certain extent, the principle of Heisenberg to the interpretation of the urban phenomenon). The instrumental means of investigation and detection, writes Muratori, have been repeatedly updated depending on ever-changing conditions, and have contributed to form "that ability to read […] in which the real production of truth of science consists" (MURATORI 1960: 13).

From reconnaissance and reconstruction of the trans-formations of the Venetian urban fabric, Muratori learns that "every structure of human operation is first of all a degree of structural connection, and only to the extent that it achieves its own individuality or, to put it more precisely, every material is but the intuition of a struc-tural behavior that takes on its degrees within its own limits of resistance and dimension" (MURATORI 1960: 7). The study of the evolution of Venetian construction sets out to clarify the role of certain constraints, of the deformability limits of materials and organisms; of the indissoluble influence, for example, "of the original foot-print, which cannot be subverted if not through com-plete destruction" (MURATORI 1960: 8). Precisely due to the resistance of the layout, "it will always be possible, unless there is complete destruction (but is it possible to completely destroy an urban layout?), to reconstruct the reality of a historical process of edification" (MURATORI 1960: 12). In the different types of urban fabric, in their respective constituent elements and the diversities of modes and treatment, it will always be possible to find the signs of the environmental-economic, politico-cultural events, the sense of transformations that are rarely radical and are often achieved with the "minimum means" of a real estate production that cannot modify itself more than partially, and case by case.

The winning project in the competition of the shoals of San Giuliano by Muratori in 1959, widely discussed and criticized, was the result – in the three proposed config-urations – of a different use and a different composition of the elements of aggregation permitted in Venetian edification. This was an interesting attempt due not to the architectural language proposed but rather to the development of the three settlement principles that take the Venetian fabric as a spatial and operative concept[25], exploring its contemporary possibilities and deformability limits.

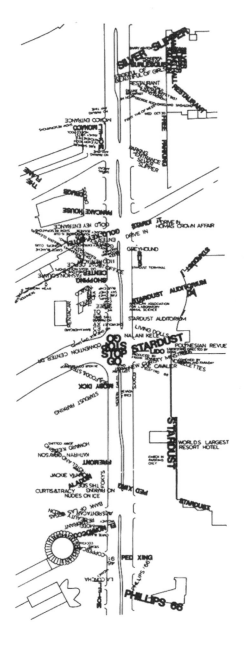

R. VENTURI, D. SCOTT BROWN, S. IZENOUR 1972, SURVEY OF THE STRIP (FROM LEARNING FROM LAS VEGAS).

How we look at things.   In 1972, Venturi, Scott Brown, and Izenour publish *Lear-ning from Las Vegas*. Again in this text, the authors are on the trail of something that has been lost: "The Forgotten Symbolism of Architectural Form", as the sub-title suggests[26]. Once again, this is an exercise of detection, surveying, and of a city-specimen, as was the case of Venice for Muratori, though now it is a specimen of urban "degeneration" from which to construct an exemplary portrait (MENDINI 1985). Along the way in the survey that observes Las Vegas as a phenomenon of architectural communication, the authors explicitly outline a different theory of design: "Learning from the existing landscape is a way of being revolutionary for an architect. Not the obvious way, which is to tear down Paris and start again, as Le Corbusier suggested in the 1920s, but another, more tolerant way; that is, to question how we look at things. The commercial strip (…) challenges the architect to take a positive, non-chip-on-the-shoulder view"[27]. Finally, again in this case, the survey is not a banal, marginal operation, since it forces the specification of innovative categories, tools, and modes of representation. In the description of Las Vegas, "Another way of understanding the new form is to describe carefully" what is there, in order to then "evolve new theories and concepts of form, more suited to twentieth-century realities […]. But how can one describe a new form and a new space using techniques based on those that came before?" (VENTURI, SCOTT BROWN, IZENOUR 1985: 28).

The new themes are the large open space, the large scale and high speed: "What is the image, or set of images, useful to an urban planner in the case of the Strip and the large and small spaces of the casinos? Which techniques – cinema, graphics or others – should be used to describe them?" (VENTURI, SCOTT BROWN, IZENOUR 1985: 66): the traditional techniques of urban planning and architecture "obstruct our comprehension of Las Vegas". What is the Las Vegas equivalent, wonder Venturi, Scott Brown, and Izenour, of the sketchbook of Roman ruins of the *Grand Tour*? A collection of images, photographs, footage, drawings, brochures, slogans, objects that go into a collage, classifications, lists. Las Vegas is not a chaotic sprawl, though "The order in this landscape is not obvious" (VENTURI, SCOTT BROWN, IZENOUR 1985: 36). It is a configuration that has not yet been understood. The pursuit of an apparently hidden structure – not immediately perceptible, above all not grasped by the collective imaginary, yet there and investigated using the tools of direct detection, recognition of systems and rules of order – is a constant concern of Venturi, Scott Brown, and Izenour, as it was previously, in a different cultural and spatial context, for Saverio Muratori.

Retroactive surveys.   In 1978, Rem Koolhaas publishes *Delirious New York*, a retroactive manifesto that describes Manhattan in terms of conjecture. The book sets out to get away from the "torrent of negative analyses" that has selected Manhattan as the "capital of perpetual crisis". It is a blueprint that shows the perfect place for a "Culture of congestion" (KOOLHAAS 1978). I will try here to observe this as a retroactive survey that investigates certain episodes of the construction of the metropolis. Koolhaas becomes a virtual detector. He also detects that which has left no trace, as in the case of the reconstruction of Dreamland or the projects that are never built (like the Grand Hotel of Gaudí, for example); he gathers studies on the maximum envelope of skyscrapers, from Hugh Ferriss to the *Regional Plan Commission*. The retroactive survey moves on different scales. It describes the island of Manhattan, but also the life of a building like the Waldorf Astoria, starting with the site, the history of its operators, and its representation in terms of plan and section.

Delirious New York contains an effort whose goal is comprehension of the Metropolis as an additive, pervasive machine whose existence has become like that of the "Nature" it has replaced: "taken for granted, almost invisible, certainly indescribable" (KOOLHAAS 1978). Actually, the description happens on multiple levels, also through projects like the *City of the Captive Globe*, an exploration of the architecture of Manhattan that formulates the questions of rapid change and the essence of the concept of the city as a legible sequence of different permanencies. The grid describes an archipelago of cities within cities: "The more each 'island' celebrates different values, the more unity of the archipelago as system is reinforced. Because 'change' is contained on the component 'islands', such a system will never have to be revised"[28].

The three cases I have proposed exist in the space between description as project and project as description. They test the limits of the deformability of fabrics, building types, and materials, by describing them; they reveal unexamined design themes through the simple act of looking; they demonstrate the constructive role of projects that have not been completed, a store of ideas revealed by retrospective detection. In each, the survey moves and becomes a source of design hypotheses; it is the starting point of a critique of the established theories and methods of design. The thesis I assert is that operations of detection and description are often connected inseparably with the proposal of new design approaches and theories and with the definition of new project territories.

[1] YOURCENAR 1936, *Comment Wang-Fô fut sauvé*, later included in the *Nouvelles orientales*.
[2] See, for example, the extended use of sampling proposed by Stefano Boeri for the construction of "eclectic atlases" during the 1990s.
[3] In his introduction to *Penser et représenter la ville* (2000), Sylvain Malfroy insists on the production of theory on the part of the different gazes aimed at the city: "[...] nous nous référerons à une conception large de la théorie au plus prés de son sens étymologique de 'méditation du spectacle des choses'. Entre la *veduta* du peintre, la photographie du témoin, le plan relief du stratège, le traité de l'utopiste, nous nous refusons à reconnaître une différence de nature profonde, au plus une différence de degré dans l'explicitation et la formalisation du propos. [...] La théorie n'est pas dans l'achèvement de la pensée mais dans le mouvement de celle-ci, dans ses innombrables recommencements à partir de questions entrevues, à partir d'aspérités du réel, effleurées à l'improviste [...]", pp. 6.
[4] For the anthology of book reviews published posthumously in 1979, Pasolini chooses the title *Descrizioni di descrizioni*, and in one of the pieces, he writes: "I have made 'descriptions'. This is all I know about my criticism. And 'descriptions' of what? Of other 'descriptions', since books are nothing else. Anthropology teaches us: there is the 'dròmenon', the fact, the thing that has occurred, the myth, and there is the 'legòmenon', its spoken description" (PASOLINI 1979).

[5] Le Corbusier, *La Charte d'Athènes avec un discours liminaire de Jean Giraudoux*, Paris, Plon, 1943. The *banlieue* is described in these terms: "C'est une sorte d'écume battant les murs de la ville" (from the edition 1957: 44).

[6] To outline a highly simplified trajectory: from *Descrivere il territorio* (II Convegno Internazionale di Urbanistica *Descrivere il territorio*, Prato, 1995) to the research *USE (Uncertain State of Europe)* published in 2003. From the conference in Prato, whose minutes have never been published and to which this book makes reference, in part, I would also like to mention the contribution of Bernardo Secchi: *Dell'utilità di descrivere ciò che si vede, si tocca, si ascolta*.

[7] See, among others: FOUCAULT 1971; WHITE 1973; 1978; SECCHI 1984.

[8] Geddes made photo-collages that suggested transformations of the state of affairs. Working on the existing image, he produced new images: "The image of the environment is made the site of our imagination, 'not only first seeing the thing as it is, but also as it may be'". DEHAENE 2002: 39. The Geddes quotation comes from *City Development: a study of parks, gardens, and culture-institutes. A report to the Dunfermline Trust*, Edinburgh: Outlook Tower, 1904, pp. 16.

[9] VIDAL DE LA BLACHE 1922, paragraph II, "Le principe de l'unité terrestre et la notion de milieu".

[10] "Les mutuelles relations de tous les organismes vivants dans un seul et même lieu, leur adaptation au milieu qui les environne" (VIDAL DE LA BLACHE 1922). The figure of Haeckel, one of the main forces in the spread of the ideas of Darwin, founder of the Monist League and a moderate support of eugenics, is a controversial one.

[11] "Des êtres hétérogènes en cohabitation et corrélation réciproque" (VIDAL DE LA BLACHE 1922: 7).

[12] "Comme les plantes se rabougrissent à défaut de chaleur et de humidité, ainsi se racornissent en pareilles conditions les groupes humains" (VIDAL DE LA BLACHE, 1922: 33-34).

[13] Adolphe Hippolyte Taine, the 19th century literary critic and historian, is known for his approach to literary history through the three categories of *race*, *milieu*, *moment*. Sainte-Beuve, while praising the method utilized in the monumental *Histoire de la Littérature anglaise*, identifies its shortcoming as the inability to explain the apparition of genius.

[14] In the introduction to the photographic reproduction (1979) of the first edition (1903) of Vidal de la Blache's *Tableau de la géographie de la France*, Claval underlines the importance of texts of agricultural geology and of description of the various French soils made possible through the production of maps that, from the middle of the 18th century, enabled more accurate knowledge of the land and structure of France.

[15] Thus in the description of Flanders, he writes of the easy communication between the *pays du sable* and the *pays nourriciers du limon*, as a balance between more or less favorable places.

[16] "Un système de lieux, dont la notion de situation (ou *Weltstellung* reprise de Ritter), est la notion-clé" (ROBIC 2000: 224).

[17] "Chaque nouveau pays est présenté comme un énigme à élucider" (PETITIER 2000: 146).

[18] "Ne s'impose au regard, elle se trahit à bien des signes" (VIDAL DE LA BLACHE 1979: 61).

[19] "Bien avant qu'au 16e siècle Guichardin écrivit que la Flandre 'était une ville continue'", writes Vidal, recalling the famous text of Ludovico Guicciardini, *Descrittione di tutti i Paesi Bassi*, published at Antwerp in 1567 (VIDAL DE LA BLACHE 1979: 79).

[20] CLAVAL and CLAVAL 1981. The two Clavals utilize the term *diffusion* to indicate a continuous and progressive increase, while *dispersion* implies fragmentation and lack of continuity.

[21] In this part of the text, I refer to the reflections contained in the talk, "Progetto come descrizione 2", given at the II Convegno Internazionale di Urbanistica *Descrivere il territorio*, Prato, 1995. Also see VIGANÒ 1999.

[22] Among others, see the research *Paris et l'agglomération parisienne* conducted by Chombart de Lauwe in 1952.

[23] The cycle starts in November of 1950 in the course on "Layout Characteristics of Buildings" in the Istituto Universitario di Architettura of Venice. The research is first published in *Palladio*, n° 3-4, 1959, and in 1960 it appears in the volume published by the Istituto Poligrafico dello Stato, to which I will make reference.

[24] "And since history is the concrete discipline par excellence, from the outset the course included field study in the form of surveys and critical reconstructions of entire neighborhoods, structure by structure, phase by phase, taking advantage of the precious experimental field offered by the historic fabric of Venice" (MURATORI 1960: 5).

[25] Various authors, "Il concorso per il quartiere residenziale alle Barene di S. Giuliano, Venezia-Mestre", in *Casabella*, n° 242, 1960; F. Purini refers to one of the project solutions of Muratori to develop his reflection on the "urban clump" as a settlement unit. See: w3.uniroma1.it/purini/testi/la%20zolla.pdf.

[26] A few years earlier, Tom Wolfe had introduced the main themes developed by Venturi, Izenour, and Scott Brown: Las Vegas was the only uniform city, together with Versailles, in the western world; in Las Vegas signs took the place of architecture: in "Las Vegas (What?) Las Vegas (Can't Hear You! Too noisy) Las Vegas!!!" (Wolfe, T., *Kandy-kolored Tangerine-Flake Streamline Baby*, New York: Farrar, Straus and Giroux, 1965).

[27] VENTURI, SCOTT BROWN, IZENOUR 1972, of the it. trad. 1985: 19. The first edition in 1972, edited by the graphic artist Muriel Cooper, was then replaced by the 1977 edition redesigned by the authors, reduced, less glamorous and costly than the first, "ugly and ordinary". See Kester Rattenbury in Rattenbury, Hardingham, 2007. The heuristic value of the maps and surveys is emphasized by the authors themselves.

[28] KOOLHAAS 1978. Also see, in part 1, chap. 2, the contribution of Ungers to the idea of the archipelago city. Koolhaas, as we know, took part in the study conducted by Ungers on Berlin, "Die Stadt in der Stadt. Berlin das Grüne Stadt Archipel" in 1977.

The concept of *porosity* comes from the natural sciences, mainly from the earth sciences and physics. It has to do with movement and resistance to movement, with phenomena of infiltration, seepage, and percolation that cross other bodies rather than a perfect void.

Porosity was used as a metaphor in 1925 by Walter Benjamin and shortly thereafter by Hermann Bloch, first to describe Naples and, more generally, Mediterranean *vivre ensemble*, which is simultaneously individual and fully shared. In its dual guise as concept and metaphor, this term is a useful tool to many people in describing and designing contemporary cities and territories, and for intercepting and revealing the major changes to which they are subject today (AMIN, THRIFT 2001; MANTIA 2005). Porosity makes reference to density and distances, taking elements of ecological rationality into account, while also having profound social implications. In a wider sense, the phantom is aroused by reflections regarding the sustainability of our actions, projects, and decisions. This complex theme, dense with ambiguities, simultaneously addresses ecological, social, and economic questions.

P. VIGANÒ, *NO VISION?*, IN DE M. MICHELIS, P. PAKESCH (EDS), *MSTAD/ MCITY, EUROPEAN CITYSCAPES*, CATALOGUE OF THE EXHIBITION- TION, KUNSTHAUS, GRAZ, 2005B, WITH S. GEERAERT, E. GIANNOTTI, A. CALÒ, N. DATTOMO, I. GUIDA (VIDEO), A. ZEIN EL DIN (PHOTO- GRAPHS). TEXT: P. VIGANÒ, 2006, *THE POROUS CITY: PROTOTIPI DI CONGLOMERATI IDIORRITMICI*, IN P. PELLEGRINI, P. VIGANÒ (EDS), *COMMENT VIVRE ENSEMBLE*, Q3, ROMA, OFFICINA, 2006.

In recent decades the European city has discovered a new kind of porosity. Connected with phenomena of large abandoned areas or of a myriad of small lots inside the urban fabric — as in the case of the city of Antwerp — this condition has brought the established city back to the center of the debate on constructed habitats. While research projects on abandoned areas have accumulated into what is now a large body of literature containing many examples, it is more difficult to grasp and describe the many small modifications that impact the small-scale tissue and transform it from within.

The experience begins with the observation and description of a new porosity that has opened up in the center of the city of Antwerp. It reflects on the tools individuals and institutions can use to approach the various grains of porosity, abandonment, and recovery, each of which requires specific strategies and techniques of infiltration.

Starting in the 1980s, two phenomena converge in Antwerp to determine widespread abandonment of the center and, more specifically, of the 19th century belt.

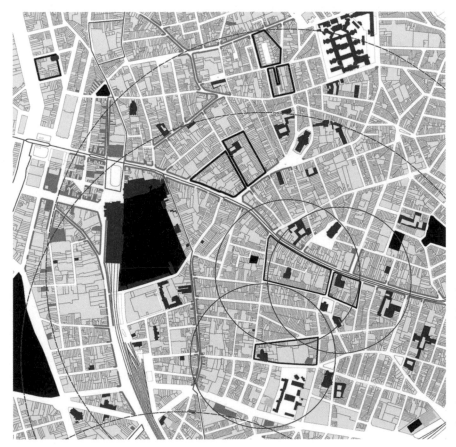

A MICRO-HISTORY: ANTWERP, REUSE OF THE BELT OF 19TH CENTURY NEIGHBORHOODS. THE CITY AS INFRASTRUCTURE:
IN BLACK, PARKS AND PLAYGROUNDS;
IN DARK RED, EQUIPMENT;
IN ORANGE, COMMERCE;
IN RED, SPORTING FACILITIES;
IN OCHRE, PLACES OF URBAN RESTRUCTURING.
THE BLACK PERIMETER INDICATES CASE STUDIES: THE RADII OF THE CIRCLES ARE 300 AND 1000 METERS.

THE "HORIZONTAL HOUSE", THE
FABRIC AND THE GREEN AREAS.

This contained a mixed fabric in which small and medium businesses coexisted with housing, often occupying the center of the block but also often facing the street. Public facilities such as schools and gymnasiums – the churches and monasteries of past centuries – could often infiltrate the long, narrow residential lots, expanding inward.

During the 1970s, the settlement model of *mixité* goes into crisis, and not only in Antwerp: in a few years that mode of coexistence of productive and residential purposes crumbles and breaks down. The consequences for the city can be dramatically seen in the policy of support for the purchase of homes, with state subsidies that permit the middle class to move into isolated houses with gardens, outside the city. Subsidies – not neighborhoods – is the policy of the Belgian government, rooted in the often described past of the "green periphery" that also continues after World War II. In those same years new populations, mostly Turkish and Moroccan, occupy large parts of the 19th century belt, therefore in the center of the city, and trigger another exodus of the Flemish population. The second phenomenon accompanies the first, and probably takes advantage of the porosity that opens up in the fabric, just as, to some extent, it determines that porosity.

In an abandoned city, where senior citizens and groups of immigrants remain, perverse dynamics are triggered: the decay of public space and buildings, the expulsion of a large part of the city from the real estate market, an increase in the gap in average income between Antwerp and its surroundings (in 1988 the gap is less than 11.9%, in 2001 it reaches 21.6% [JANSSENS, DE WAEL 2005: 140]). Phenomena of this kind have been described in situations of shrinking cities; these are cities where sudden exodus and abandonment have happened, often in the wake of major economic transformations.

The main economic activity of the city, as the second largest European port for oil, does not stop; the port slowly but surely moves north, outside the city, leaving large abandoned areas in its wake: railway platforms, basins, docks, depots, warehouses. This creates large-grain porosity. It gets harder to live in the city, which becomes less and less appealing to new inhabitants capable of contributing to its economic profile.

In this situation of crisis, the first attempts to recover the large abandoned areas do not catch on, and only in recent years have we seen the start of strategic thinking about the potential transformation of the city. A state of abandonment prevails inside the 19th century belt, and this constitutes the starting point of a new porosity and availability to be interpreted. One of the main elements that deter-

THE HORIZONTAL HOUSE, THE VERTICAL HOUSE: ISOMETRICS OF THE HOUSES INSERTED IN THEIR CONTEXTS. STILLS FROM THE VIDEOS.

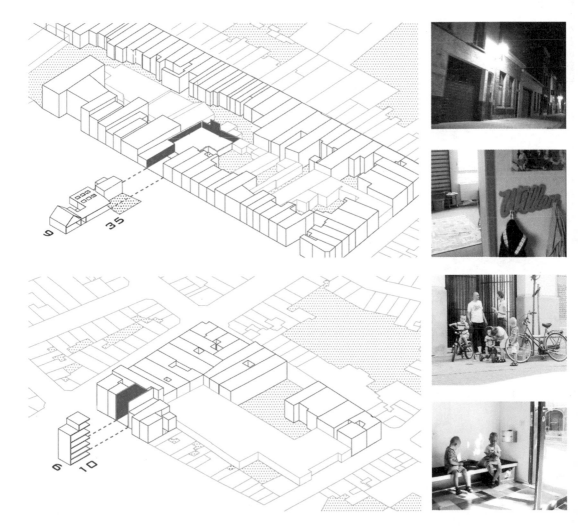

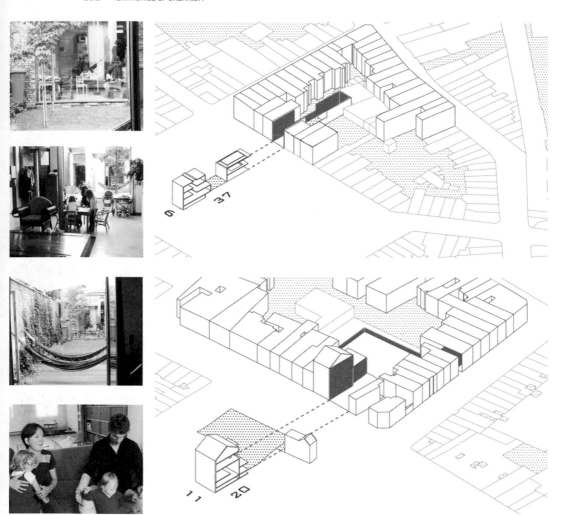

THE DOUBLE HOUSE, THE HOUSE OF STACKED HOUSES: ISOMETRICS OF THE HOUSES INSERTED IN THEIR CONTEXTS. STILLS FROM THE VIDEOS.

mines the porosity is the small size of the lots and the properties (VIGANÒ 2005b). The zone resists initial public renewal efforts that attempt to make the city more attractive for investors; on the other hand, it is open, by its sizing and nature, to strategies of individual access – projects on the scale of the family, of the young couple.

The choice of returning to live inside the dense fabric of the city, of coming to terms with new types of coexistence, but also of remaining connected to that large infrastructural platform that is the city: these are the individual decisions that trigger profound modifications of the urban fabric, outside the ideological or nostalgic choices regarding the space of the *vivre ensemble* of the past. In the case of Antwerp, the lots welcome both a contemporary idea of dwelling and the space of achievement of the dream connected with living: in the city, but in large homes with gardens, possessing the desired privacy and comfort.

Standards or prototypes? The Belgian standard of living is very high if compared, for example, to that of Holland or France, and the welfare model corresponds to this, precisely because of the centrality of the individual. It is the result of a long-term tradition in which the fundamental characteristics of inhabitable space remain stable: open living in the solid city.

In 2003 Antwerp was ranked 38th among 40 Belgian cities for quality of public spaces and their maintenance (*Stadsmonografie Antwerpen*, 2003). However with many new projects for public spaces across the city, intersecting with or bordering on conflicted, compromised places in need of signals of renewal, "subtle porosities" emerge from different and fragmented individual initiatives that nevertheless participate in the construction of a collective and public capital such as the city.

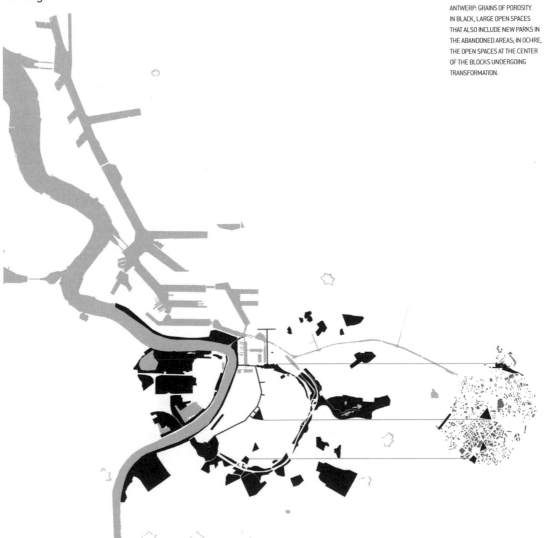

ANTWERP: GRAINS OF POROSITY.
IN BLACK, LARGE OPEN SPACES
THAT ALSO INCLUDE NEW PARKS IN
THE ABANDONED AREAS; IN OCHRE,
THE OPEN SPACES AT THE CENTER
OF THE BLOCKS UNDERGOING
TRANSFORMATION.

THE HORIZONTAL HOUSE: MODEL.

Tools for interpreting porosity

Interpreting porosity calls for conceptual and material tools. During the research on the subtle porosities of Antwerp, situations have been described by attempting to shed light on the relationship between persons and things. Uncertainty – for example, going back to live in a city where one is now part of a minority – is approached by relying on objects to construct possible forms of order. Objects are consolidated, attaching them to constructed orders. The world is put to the test by this disquiet, pointing to the value of objects from which an adjusted order can emerge.

The construction of micro-histories had to contain close analysis of the physical dimensions of space: the dimension of a dwelling, a patio, a living room, inside the urban fabric; the measuring of the distance from the daycare center, the school, shops, work; the degree of unfinished self-construction, of the old garage that is transformed into a home, one step at a time. To illustrate certain hypotheses it was essential to use maps and construct large models, take photographs and make videos, record interviews and stories, and visit places.

The first hypothesis: that the return to living in the city is not driven by a nostalgic idea of community, and that such romantic hindrances would not stand up to the harsh reality of the *vivre ensemble*: because coexistence does not always construct a situation (BOLTANSKI, THÉVENOT 1991: 51).

The second hypothesis: that only the possibility of finding idiorhythmic spaces (BARTHES 2002) inside the urban conglomerate could attract young Flemish families into city parts where the majority of the population speaks a non-European language and belongs to different cultures.

The third hypothesis: that the great resource of the European city is that it is still a vast, equipped platform. It is from this standpoint that the question of proximity and of the *vivre ensemble* becomes crucial. The instruments utilized and the operations implemented were therefore closely oriented by a group of hypotheses and concepts.

THE HORIZONTAL HOUSE: MODEL.

Porosity of material and porosity of fracture

Porosity can be subdivided into two large families, with the first referring to the porous nature of material, and the second to a porosity that is the result of traumatic events.

The first family has to do with the structure of a material, with the way it is designed, and its capacity to be porous. Designing porosity takes on a meaning pertaining to the forms of space and the devices that can make it open to different individual rhythms. Often the porosity opens up at the end of an economic or social cycle. This is the second family, arising in the wake of a fracture in the modes of use of space, making it possible to rethink the city, starting with the extension and importance of the fracture that has taken place. It is very difficult to foresee this type of porosity, but we at least have to know how to interpret and recognize it as project material.

Porosity and individualization

The case of the subtle porosities of Antwerp is interesting because they reveal the breakdown of a balance among different functions, in lifestyles, while at the same time demonstrate the functioning of the fabric as an idiorhythmic contemporary conglomerate. This permits a reading of the process of individualization through urban space: "a historical process of individualization, as progressive tension towards the achievement of a substantial freedom of the individual" (Paci 2005) and a fundamental part of western modernity.

B. SECCHI, P. VIGANÒ, STRUCTURAL PLAN OF ANTWERP, 2003–2006 (EXTERNAL TEAM) WITH M. BALLARIN, A. CALÒ, N. DATTOMO, F. VANIN (TO THE RIGHT). THE POROUS CITY: GUIDELINES. POROSITY: THE DENSE CITY (THE 19TH CENTURY BELT, THE WORKERS' DISTRICTS, SOME PARTS FROM THE START OF THE 20TH CENTURY) THE MORPHOLOGY OF THESE AREAS CONSISTS IN INTERNAL GARDENS, SQUARES, ROUTES, AND BLOCKS THAT COULD BECOME MORE ACCESSIBLE AND BETTER INTEGRATED WITH THE REST OF THE URBAN CONTEXT.

POROSITY: THE MODERN CITY. THE THEME IS THAT OF THE PUBLIC SPACE BETWEEN BUILDINGS. A NEW DESIGN OF THE IN-BETWEEN SPACE AND INTRODUCTION OF NEW ACTIVITIES IN THE EXISTING FABRIC.

POROSITY: THE GARDEN CITY. THE PATTERN OF LOW-DENSITY HOUSING, TOGETHER WITH PATCHES OF EXISTING NATURALNESS, CAN BE TRANSFORMED INTO A CONTINUOUS – THOUGH FRAGMENTED – OPEN SPACE THAT IS ACCESSIBLE AND PUBLIC.

In the two types of porosity outlined above, the capacity to absorb practices and transformations of different natures — both individual and collective — is a fundamental question. Also in the large grain of the large urban park, the project has to introduce spatial wrinkles, folds of meaning; it has to make itself "porous" to absorb individual forms of conduct and collective behaviors. This idea of substantial and positive freedom of the individual is not opposed to the idea of social cohesion, although it is opposed to "the model of community organization [...] belonging to tribalism or to the collectivism of totalitarian systems" (GERMANI 1991:23). The concept of individualization, having fewer ideological connotations than that of individualism, is a structural process produced by increasing social articulation and fragmentation. If the society is to be the place of realization of the individual and not just the place that guarantees his or her security, the forms of space porous to individuals and their practices permit reflection on their role in the construction of a common asset.

Porosity and possibility

From what has been explored above, there is a clear relationship between the concept of porosity and a particular strategy of attention: to look between and inside the folds, not only with the analytical gaze of the explorer or the archaeologist, not only in search of signals of change, but also within a specific design strategy that rests on the conviction of the role of the individual in the transformation of space; that investigates how these changes can be inserted within a common strategy, then reinterpreted, directed, and utilized. The materials of the project that come to terms with the concept of porosity are the situation, the configuration, the terrain, and the possibilities that are wedged into them, without characteristics of necessity and environmental determinism, as Vidal de la Blache believed, that are partially bordered by the choices already made, by their stratification, and inertia.
The idea of *idiographie* developed by Vidal's school belongs to the analysis of porosity. Idiographies and idiorhythms address what is unique and specific.

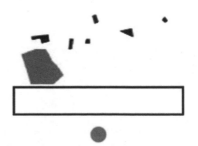

POROSITY: HETEROGENEOUS AREAS. THE PATCHWORK OF DIFFERENT DENSITIES AND ELEMENTS CAN OFFER AN UNEXPECTED OPPORTUNITY TO CREATE NEW OPEN SPACES, ALONG WITH NEW ACCESSIBILITY FOR ACTIVITIES AND PRACTICES (TO THE LEFT).

POROSITY: THE PORT. THE EXISTING OPEN AREAS, NOT UTILIZED, CAN BE SEEN AS SPACES THAT CAN BE INFILTRATED BY NATURE. THE NEW GREEN PATCHES CAN TEMPORARILY OCCUPY THE SITES IN KEEPING WITH THE MOVEMENTS OF THE PORT.

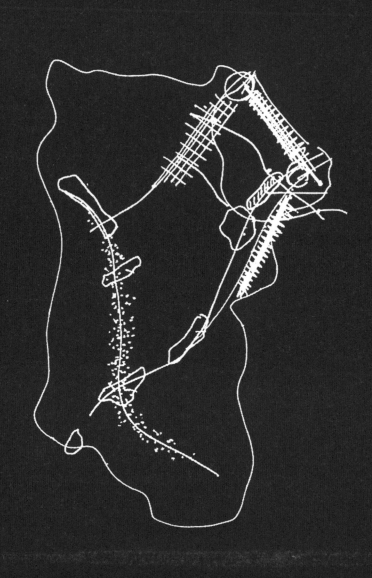

The project that makes full use of its descriptive dimension moves inside a specific cognitive strategy: the apprehension and learning of a place happens using the tools of design and its techniques.

Images as grooves

The project as description feeds on vague images that cannot be defined in detail except by means of new descriptions, which in turn are forcefully oriented by the utilized interpretative image itself. The importance of the development of images emerges as a constant from design reflections on a large scale: the image is the intermediary between the physical and conceptual territories of the project. Observing certain explorations on a large scale over the last few decades[1], I can recognize common characteristics and differences. I will start with the differences.

The letter π.   In the territorial plan of Pescara (SECCHI, BIANCHETTI, VIGANÒ 1998), the image of the Greek letter π summed up the major features of the territorial structure, the two main valleys and the coast, while at the same time selecting the strategic space in which to position the main transformation projects. This large figure offered an iconic way of conveying the main territorial forms, as well as the major mobility infrastructures (railroad lines, toll roads, and highways) and the focal points of the local economy – the large industrial areas along rivers, the coastal city of services and tourism. A description and design figure in its own right, standing out against a background, the π left a large part of the territory indefinite, without form, so its design had to be found in the relationships this area could establish with the strong structure of the two valleys and the coast. During the development of the plan, the image of the letter π, though it guided certain important design choices, became less forceful, though it was never entirely disrupted or replaced.

The reticular city.   In the territorial plan of the province of La Spezia[2], the *reticular city* is the image of a way of generating low-density territory with an open structure in which relationships between *full* and *empty* are inverted, and a distinction between city and country can no longer be discerned. The net figure intersects the polycentric pattern without either of the two managing to prevail; it organizes a

B. SECCHI, 1998, SKETCH OF THE GREEK "π", FROM SECCHI, BIANCHETTI, VIGANÒ, 1998, TERRITORIAL PLAN OF THE PROVINCE OF PESCARA.

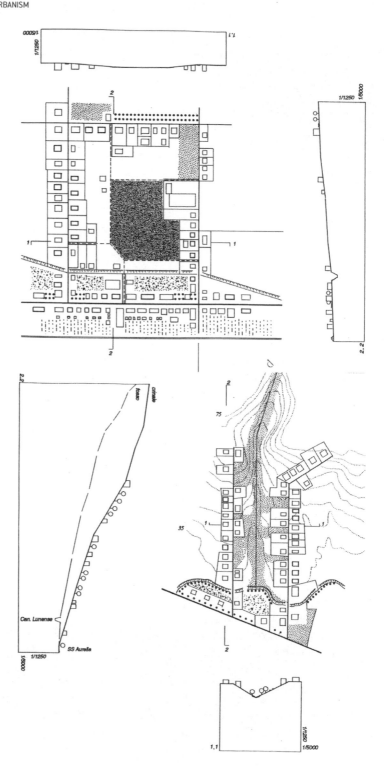

B. SECCHI, A. LANZANI, P. VIGANÒ,
TERRITORIAL COORDINATION PLAN
OF LA SPEZIA AND VAL DI MAGRA,
1993, WITH P. CASTIGNANI, D.
FORMENTINI, L. PIPERNO, G.
TRENTUNO.

SETTLEMENT TYPE SCHEMES: THE
BRACKET AND ITS COMPLETION;
THE HILLSIDE "COMB".
THE URBANIZED COUNTRYSIDE:
INTERPRETATIVE SKETCHES OF
THE LARGE IMAGE PROPOSED
BY G. SAMONÀ IN THE PLAN FOR
TRENTINO.

CENTRAL AXIS
VEGETABLE GARDENS
IN-BETWEEN AGRICULTURAL AREAS
COLLECTIVE GREEN SPACES
ROUTES
TREES
FOREST
ADDED BUILDINGS

minimum vocabulary of elements, deploying houses on lots, the canal, vegetation on the banks and scattered houses as materials with which to construct a new habitat.

The reticular city is the image of a mode of territorial construction. It is a procedural image because it describes the ways in which urbanization has developed, with a certain amount of repetition, leaving large gaps inside an inhabited grid structured by the long-term signs of waterways and streets. It is the image of a process, because the conclusive form (the form, that is, that appeared to be the result of the rules described above) would not be fully represented in the plan. Certain rules, only partially indicated by the project, interpreted an already existing logic, making it visible.

The choice of this image requires at least two conditions: an evaluation of its rationality and capacity to generate spaces that are not functionally separated, equipped with greater quality than other spaces based on zoning, or on the non-selective insertion in large cultivated plots of hybrid spaces of agriculture, production, and residence. The rationality of this settlement mode seemed to be minimal, connected with the use and reuse of an infrastructural support already existing, to a large extent, and with the historic subdivisions of the territory (centuriation). The character was very similar to that of many other places in Europe, not only when marked by the Roman *aggeratio*. All the territories invested with large-scale fixed capital and enormous quantities of dead labor constructed conditions of widespread inhabitability over the long term. The image of the reticular city made this condition visible – a condition that may not be exceptional, but whose importance and extension are often underestimated[3].

The project for the territory of the Val di Magra brings recognition to a habitat with very different characteristics from those of the compact city, expressed in tables that provide the settlement rules: the linear settlement and the "quad" on the plains, the hillside perpendicular array or at the foot of the hills, for example. These are rules that describe structural characteristics in the contemporary space of the Magra valley. The unitary character of the project is determined only by the relationships, the frame that covers a number of themes: the market-street, the high center, the reticular city, the river. In the "Piana" between Florence, Prato, and Pistoia, in ways that are similar, in many respects, domino and puzzle seemed to constitute the two fields of rules that combined procedures (the sequence of moves of a game of dominoes) and the pursuit of a figure inscribed in the territory (the puzzle that is composed in the attempt to reconstruct a figure that has been broken into pieces), in this case its fundamental ecological structure (VIGANÒ 1999).

The isotropic territory.   In the latest research on the metropolitan area of Venice[4] ,the image of an isotropic territory indicates certain physical properties the project sets out to reinforce after having recognized them as long-term characteristics. The idea that the direction can be of little influence in the construction of a territorial project is extreme and radical, capable of subjecting to critique all the conceptual baggage available to us today, from *urbanisme de l'axe* to pole-based planning. However, this has not prevented it from being taken into consideration over time. Direction is the "way along which persons or things move", "the address, the route, the trend"; but it is also an "act, the effect of directing, of commanding"[5]. The design of isotropy is useful to understand the distance between the concrete ways in which the territory around Venice has taken form along with the ideas and projects that attempt to change its characteristics. They superimpose rankings and functional and social separations on a space densely infrastructured over the long term in opposing ways.

The territory as park.   In the plan for the province of Lecce[6] finally, the image of Salento as a park comes from immersion in its different landscapes and orients the entire construction of the territorial project. This is a synthetic image in which the observations and types of knowledge that rotate around the territory can interact. It is blurry enough to be premature, that is it can guide research in greater depth to reinforce or refute the image. "Salento as park" is not an image of form, since multiple types of parks exist; it does not contain the rules of the game or the formal tenets of its implementation, like the procedural/process-based image of the reticular city. It is not an image built starting with physical properties, like that of the isotropic territory. The park is an image of status, an image in the proper sense of the term, that evokes an atmosphere that is an alternative to that of the city, though cities are contained inside it. It is perhaps the image that asserts with greatest clarity the inversion of the relations between open and built space in the dispersed and fragmented city, putting the design of the void at the center of the project, no longer making it a neutral backdrop against which to insert infrastructure and settlement, but rather a fundamental object of research and transformation. The park, on the scale of an entire territory, becomes a settlement model in its own right, to virtually place alongside the "three human settlements" proposed by Le Corbusier: the compact city, the linear industrial city, and the rural nucleus. It is the arrival point of lengthy design reflection that marks the history of modern urbanism and that accompanies, like a parallel path, the most outstanding episodes of the construction of the city of plenitudes.

The elements shared by the various images lie in the role they play for interpretation and design, in their capacity to structure the gaze that reads, and to already be design. They lie in the hybrid space that they open between description and project, in their capacity to unify disparate gazes hailing from the multiplicity of disciplines involved in the construction of a territorial project, in their ability to supply a vanishing point for the different trajectories that they trace. What unites the different images is the attempt to disrupt the traditional sequences of construction of the territorial project and, at the same time, to propose different sequences of urban spaces and materials. Attempting to say what they are and

what they could become establishes a tense and conflicted relationship, but one with depth, between the collective imaginary and that
of the discipline, between territories and project.

Images of the city-territory

The long Italian tradition of thinking on the large scale and the new
urban dimension has not been without influence on the works outlined above and on their different images, starting with the powerful vision of the "urbanized countryside" contained in the Trentino
Plan of Giuseppe Samonà[7], an image many erroneously interpret
as being exclusively analytical, almost an oxymoron. The reading
of the plan and texts of Samonà clarifies the design meaning. The
work proposed a new settlement form that integrated the countryside and the Alpine slopes, giving them different roles than in the
past, inside a new condition of widespread urbanization, distributed
in egalitarian ways over a territory.

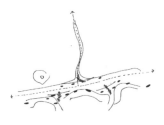

Recognition of settlement dispersion as a long-term characteristic
is not immediate in the research that expands in Italy from the late
1950s to the early 1960s. What is urgent is to understand the changes
in progress, interpreted as original and recent, to impact and modify
the characteristics of the city. A new urban dimension, a new scale (of
design), and a new society (much more mobile than in the past) are
the three questions that orient the multitudes of research conducted
in Italy on territorial transformations starting at the end of the 1950s.
Samonà is a great nurturer of images that are intriguingly located

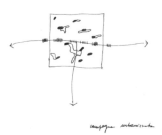

on the borderline between interpretation and prediction. These images simultaneously describe and indicate a direction for the construction of a new type of space.
The image of the urbanized countryside is one of the most powerful for indicating
a new territorial scale of the project. It comes from a new awareness of the crisis of
the countryside, the risk of disappearance of the traditional scattered settlements
of the Adige Valley, and it announces the territorial project driven by all this, recon-
ceptualizing the urbanized countryside as a possible settlement form.

THE URBANIZED COUNTRYSIDE:
INTERPRETATIVE SKETCHES OF
THE LARGE IMAGE PROPOSED
BY G. SAMONÀ IN THE TRENTINO
PLAIN.

At the conference in Stresa on the new urban dimension (organized by Ilses in
1962), other large interpretative images contain reflections on changing lifestyles
and modes of use of the territory. De Carlo picks up Geddes' and Mumford's image
of the city-region and links it to increased affluence, the acceleration of social and
territorial mobility, and the resulting multiplication of the possibilities of choice.
Later, in the inter-municipal plan of Milan, De Carlo uses the image of the urban
continuum, in which urban planning has to identify systems and structures. The
structure of the urban form is the theme that interests him, and particularly the
structure of a new form of urban-ness that is scattered and open.

The image of the city-region introduced by Geddes, together with the newly coined term "conurbation" in 1915, described a new spatial entity, a community, a specific urban federation of the English territory that represented its lifestyles: a city-region, a community that lives in an extended, polynuclear, and almost entirely urbanized space. But this clear definition of *territory* in which the city no longer has a fundamental generating role is soon joined by other meanings and attributions, to the point of contradicting the original formulation. In the 1920s, Lewis Mumford and Thomas Adams discuss, regarding New York, the terms of the "regional city", and for both of them the specter to be remedied is that of the congested and polluted industrial city, as opposed to the contents that fill the image of the "city-region" (Robic 1998)[8]. For Mumford this takes the form of a communitarian ideology that is represented in the project of a network of satellite cities; for Adams it is instead the expression of a logic of reinforcement of the role of the city, of the large city as the center of a vast region. Communitarianism and metropolitanism, though applying similar instruments (such as expansion through new units) diverge deeply, and in time a fundamental ambiguity arises and takes root in the image of the city-region.

Also in 1962, Piccinato, Quilici, and Tafuri, representing the Roman studio AUA, introduce – in *Casabella Continuità* and in the Italian debate – the image of the "*città territorio*" that indicates not only a new vantage point, but once again also a shift of scale. The incapacity of urbanism and planning to interpret the new phenomena is seen as eminently ideological, with the latter engaged in elaborating "microcosms, 'neighborhoods' that are absolutely detached from the surrounding environment in all its aspects" (Piccinato, Quilici, Tafuri 1962: 16). Advancing a strong critique of the intimist solutions of neo-empiricism, the authors point not to a problem of tools of intervention or investigation, but to one of identification of themes and their relationships, which would require the development of a new "structural pattern". The AUA[9] group, like De Carlo, starts with the consideration of the raising of the level of life, but introduces a new focus on the theme of free time – the spread of *loisirs*, the rise of the second home, the role of television – and the gradual dispersion of production that "evaporates" in the territory and abandons its proximity to the urban center. The accent is on the acceleration of transformations, on territorial dynamics, the mutations of ways of living together. These positions insist on the risk of a lack of comprehension of the transformations in progress, on the need for new techniques of investigation, new readings and survey, and on the urgent need to reformulate the tools of the project, a question that is much more serious than that of the ageing of institutional and legislative measures.

Around the image of the "city-territory", an idea of the project as process takes form, which no longer has the possibility of approaching the whole, and where the individual exercises freedom and autonomy: "The city-territory shifts its field of application from total urban planning to the identification of points on which to exert leverage [...] though this certainly does not imply rejection of urban plan-

TERRITORIES OF SETTLEMENT DISPERSION: VIEW OF THE PLAIN BETWEEN CASTELFRANCO VENETO AND BASSANO.

ning on a territorial scale" (PICCINATO, QUILICI, TAFURI 1962: 17). The dilation of the scale of the urban phenomenon does not imply more widespread planning; rather, it implies the selection of places against the backdrop of an urban project on a large scale. During this same period the construction of the Trentino plan on the part of Samonà was beginning. It marked the first territorial and development plan in Italy prepared by urbanists and not just economists.

B. SECCHI, P. VIGANÒ, A PLAN FOR CASSOLA (VICENZA), 2005, WITH U. DEGLI UBERTI AND T. LOMBARDO. "THE RED AND THE BLACK" (IN RED, WHAT HAS BEEN ADDED TO THE EXISTING SETTLEMENT STRUCTURE OVER THE LAST 40 YEARS).

THE ENVIRONMENTAL SYSTEM: IN RED, THE WATERWAYS; IN DARK GRAY, EXISTING AND PLANNED HEDGES; IN LIGHT GRAY, RESERVES OF PERMEABILITY; IN BLACK, BICYCLE PATHS AND WALKWAYS.

The territory described in those years is simultaneously the result and the motor of important changes in society (ARDIGÒ 1967). The problem consists in giving a "democratic direction" to such potentialities to meet the demand of the new affluent society "to rediscover, in every condition, the variety of contacts and choices the city offers" (PICCINATO, QUILICI, TAFURI 1962: 17). The idea of a movement in the direction of "more highly evolved forms of territorial organization" is shared by many, though it is not clear whether or not these are socially positive forms. Unlike the Modern Movement, which had proposed and overlaid a model of progressive development, AUA underlines the impossibility of maintaining this ambition, evoking the crisis of a constructivist position and proposing a way of getting beyond a rationalist attitude, to "instead draw on a continuous process of rationalization". The possible emergence of a new form of associated life and a new territory should be kept in a state of observation and subjected to constant verification.

If we compare all this (BARATTUCCI 2004) to the thinking – in the same years or those immediately following – within the French debate, we can see certain important differences in terminology. In the second half of the 1960s, Henri Lefebvre, in the introduction to the study on the *habitat pavillonnaire*, speaks of the *ville* éclatée: the accent is on the fragmentary character of contemporary space, as if after an explosion (LEFEBVRE 1966). In 1976, Bauer and Roux coin the term *rurbanisation*, an analytical and conceptual tool with which to approach a city that extends, a *ville* éparpillée, where the focus on the fragment and the loss of unity gives way to a problem of confused and spread relationships between different settlement materials. The question is no longer that of loss, but of finding a new urban dimension, new relationships between territorial spaces. As in the case of the urbanized countryside of Samonà, what emerges is the need to imag-

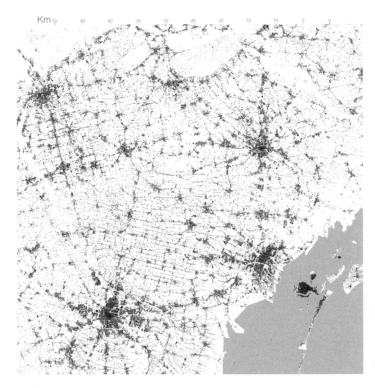

WATER AND ASPHALT, THE
DESIGN OF ISOTROPY IN THE
METROPOLITAN AREA OF VENICE:
IN GRAY, THE CONSTRUCTED
FABRIC; IN RED, THE NATURAL
AND ARTIFICIAL WATER
NETWORKS.

THE RELATIONSHIP BETWEEN
THE STRUCTURE OF THE WATER
AND THE SETTLEMENTS IS
STRIKING (CARTA TECNICA
REGIONE VENETO, 2007).

ine an updated relationship with agricultural areas, a question that, even forty years later, despite the attempt to institutionalize new urban forms, remains fraught with difficulties.

The accent placed on the city-territory weakens in the successive descriptive images that seem to portray a changed condition rather than constructing a new one. The theme of rurality becomes marginalized, swallowed up by the traditional concerns of urban planning regarding consumption of land area. The studies intensify on fragmented growth that leads to widespread, dispersed urbanization, which seems, in certain areas, to announce a stronger metropolitan integration[10]. Morphological situations, settlement principles, and social conditions are investigated[11] on different scales, focusing on the elements of *longue durée* that, while they are often negated by settlement dynamics, seem to be the only factors capable of guiding the new territorial organizations. Therefore it is during the course of the 1980s, parallel to renewed reflection on time and history, that observation begins of the relationships between new forms and long-term territorial characteristics. Perhaps this is what shifts the vantage point from the city and the theme of its new dimension to the territory as a place of long-term rationality – and thus the fundamental base of study for research on new structures. Together with the reinforcement of the themes of ecology and long-term factors, images and design themes emerge that identify the territory as the crucial space of the contemporary project.

At the start of the 1990s, the image of the *diffuse city* is introduced by Francesco Indovina and Bernardo Secchi; it is a term that describes a type of spatial organization marked by the presence of certain urban characteristics as opposed to others. The *diffuse city* describes not only the consequences of the dispersion of housing but also of all urban activities. It is the result of spontaneous actions and policies: a first wave connected with improvements in the living conditions of the agricultural population as it shifts gradually into the secondary sector, followed by a second wave generated by the abandonment of the city on the part of its inhabitants (middle classes, dissatisfaction with life in the city), attracted by lower housing costs and the possibility of living in a different way. The diffuse city is an interpretative image that makes it possible to address the problem of individual and collective freedom, which is crucial for Indovina, and to formulate a judgment of the new phenomenon. Unlike a traditional metropolitan area marked by vertical connections and a strong hierarchy, the diffuse city is woven out of horizontal relationships and has a weaker system of ranked relations. Inside this horizontal territory, which Secchi begins to describe on a European scale, we can recognize filaments, platforms, points of density that have not yet been stabilized, gaps waiting to be filled – a world of objects. Secchi suggests, echoing Banham (SECCHI 1991), a new ecology, or even multiple ecologies linked among other things to the different histories of the urban phenomenon in Europe, waiting to be described and projected into an overall reflection that recognizes their themes[12]. While the expression *diffuse city* soon becomes an international point of reference, it contributes to a new local political awareness that asserts its own identity, in opposition to that of the traditional city. In this case, the territory of the project is political.

[1] In this part I will discuss some of the territorial design explorations conducted over the last twenty years with Bernardo Secchi.

[2] Secchi, Lanzani, Viganò, *Piano Territoriale di Coordinamento di La Spezia e Val di Magra*, 1993; SECCHI 1994.

[3] Today, for example, the expansion of the Chinese megacities depletes enormous agricultural areas infrastructured by the water network and routes, instead of concentrating on a few places of higher density. The interesting point, on the other hand, is that they could represent the imprint of an innovative settlement model from the viewpoint of energy and, more generally, of ecology.

[4] See: *Territory 1, The Project of Isotrophy*.

[5] *Il nuovo Zingarelli, Vocabolario della lingua italiana*, Bologna, Zanichelli, 1988.

[6] See: *Territory 2, Conceptual Shifts*.

[7] The construction of the plan of the Trentino, by Giuseppe Samonà, begins in the early 1960s and comes to terms with the theme of the large territorial scale, the form of the territory, and the construction of a project (Provincia autonoma di Trento, 1968, *Piano urbanistico del Trentino*, Padova: Marsilio).

[8] Another widespread text was that of Robert E. Dickinson (1947), *City Region and Regionalism*. In this case, *City-Region* means: an "area of functional association with the city" (DICKINSON 1964: 227).

[9] AUA (Studio Architetti Urbanisti Associati) was founded in 1961, and its manifesto was signed by M. Tafuri, L. Barbera, B. Rossi Doria, S. Ray, M. Teodori, and later by E. Fattinnanzi. The projects of the group were often done together with other professionals, including V. Quilici, S. Bracco, A. Calza Bini, M. La Perna, C. Maroni, G. Moneta, among others.

[10] G. Astengo, research *It. Urb.* case study of the Brenta River in the central Veneto area.

[11] PICCINATO, DE LUCA 1983; SARTORE 1988; PICCINATO, SARTORE 1990.

[12] In the 1990s, Boeri, Lanzani, and Marini, in *Il territorio che cambia* (1993), try to hold together the various dimensions of change revealed by the crisis of traditional and natural morphogenetic elements, in pursuit of a new geographical image. The focus on persisting factors and the value of tracings is joined by the attempt to describe the different modes of change in the territory, observing the practices without disjoining them from the spaces. The research Itaten (CLEMENTI, DE MATTEIS, PALERMO 1996) attempts to report on the transformations of the Italian territory. See also: ZARDINI 1996; MUNARIN, TOSI 2001; VIGANÒ 2001; BIANCHETTI 2004; VIGANÒ 2004. It is also important to mention the research of N. Portas on the territories of dispersion in Portugal, of A. Font on the metropolitan region of Barcelona, the urban landscape of T. Sieverts, the research of M. Smets on settlement dispersion in Flanders, and the *Atlas*, edited by B. De Meulder and M. Dehaene for a part of the same territory; more recently, DIENER 2006.

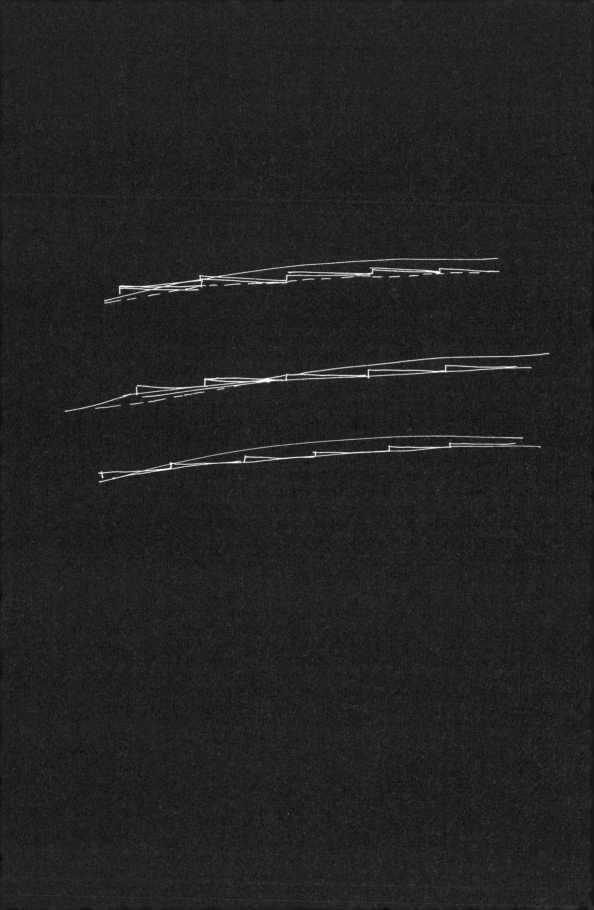

Every act of description is an attempt at logical reconstruction of the world, pro-ceeding through deconstructions, erasures, and highlighting. Several years ago I attempted to describe certain characteristics of the contemporary habitat by using a limited number of recent projects. I will return to the main parts of that text (VIGANÒ 1994) without adding new references, because the purpose is not to supply an overview of descriptive projects, but to clarify certain modes of descrip-tion that are possible through design. In particular, this will allow me to introduce the hypothesis of design as description of the structural attributes of the contem-porary environment. The thesis is that descriptive and representative projects are those that do not deny and do not claim to change the structural attributes of contemporary space, but instead reveal the elements of rationality contained in those very characteristics. Starting with the physical and material constitution of the territory, they can reference practices and economies that are endowed, in turn, with different forms of rationality.

If we set out to indicate the theme or, at times, the argument that has most force-fully marked the design debate in recent decades, we have to point to the rela-tionship with the context. The context is an unavoidable parameter of interaction in terms of acceptance or separation, agreement or rejection. It has been placed alongside and in opposition to the claims of autonomy of design choices, imply-ing design's independence from a given context and its supernational nature. It is clearly a theme that has been extensively addressed and has imposed a recog-nizable direction in the urban planning debate from the 1950s to the present, as reflected by one of the most fertile beginnings of this discussion, in the issue of *Edilizia moderna* on the form of the territory (GREGOTTI 1966).
During the 1980s, references to context, with the development of a theory of modification[1], marked the rise and assertion of issues connected to the need to take a position inside a given situation determined, to a great extent, by existing circumstances. However, even a rhetorical use of the appeal to context should not mislead us. What has been designed and produced has instead pointed to a gap between an abstract position and design practice that is often quite distant from that position. Self-referential posturing, descriptivism, mimetics, and problem solving[2] are attitudes that differ significantly from those we would like to exam-ine here. It has only been recently, after years of attempts at description, that

CEMETERY OF KORTRIJK (BELGIUM): THREE WAYS OF DESCRIBING A SLOPE.

certain projects have reinforced their descriptive dimension in order to be able to intercept and explicitly account for the structural connotations of contemporary space. These are projects that subject their rules and the mechanisms of their construction to the description of the place, the time, and the society in which they exist. Besides coinciding with this initial definition, the projects illustrated have at least one other factor in common: they spring from the irregularities of the world, without presuming the existence of a perfectly homogeneous situation. Anything deviating from, "If the world were totally regular and homogeneous, there would be no forces, and no forms"[3], and these projects would certainly not exist. They describe — i.e. they represent and make visible, perceptible, and evident — the irregularities of the world or of its image.

Landforms

The project for the expansion of the southern part of the city of Kortrijk, Hoog Kortrijk[4], develops a system of discontinuous collective urban spaces that are not in proximity to each other. It juxtaposes new functions on fragments of cultivated territory: the office park, new central places, the shopping center, themes that describe practices of contemporary use without setting precise limits of pertinence. The project describes the form of the territory, its waves, its peaks and valleys. It is not an imitative description that sets out to organize a reasoned path to a place, introducing a "difference" that selects the elements in a precise way. The large *galettes*, horizontal areas that reveal the movements of the terrain, are an attempt to update the vocabulary of contemporary public space, making it cope with important themes, such as the expanses of parking required near large attractions outside the compact city; the distances between objects, the fair, commerce, but also the presence of vegetable masses; and the woods beside a large traffic junction the new space absorbs along its way. The *galettes* are "landforms" in the sense of an architecture of specific topographical relations. They are horizontal forms of order that orient the choice of the settlement principle of the objects.

The project for the cemetery and park of Hoog Kortrijk proposes reflection on the nature of public space and the role of the form of the territory in the design of the contemporary city. The project describes the shape of the terrain in great detail, traversing and identifying it. Placed on a slope outside the established center and along the axis that crosses Kortrijk from north to south, the cemetery is a path that descends from the crest all the way to the base of the valley. The crest marks the natural limits of a rarefied city, beyond which the view of the countryside opens out.

The project builds its knowledge of the place through three main moves — three sections whose differences highlight the shape of the terrain and therefore its morphological and geographical characteristics. The first section rests on the existing levels and follows them without modification. It has an irregular slope; it defines the main route that descends to the base of the valley, flanked by a

B. SECCHI, P. VIGANÒ, PROJECT FOR THE CEMETERY AND PARK OF KORTRIJK (WITH A. SECCHI, B. MARTINO, PRELIMINARY PROJECT; N. GOFFI, G. MANZONI, B. MARTINO, G. COMANA, G. GIUNTA, DEFINITIVE PROJECT), COMPLETED IN 2000. VIEW OF THE PLATEAUX FROM THE STREET AT THE CREST, TOWARDS THE VALLEY.

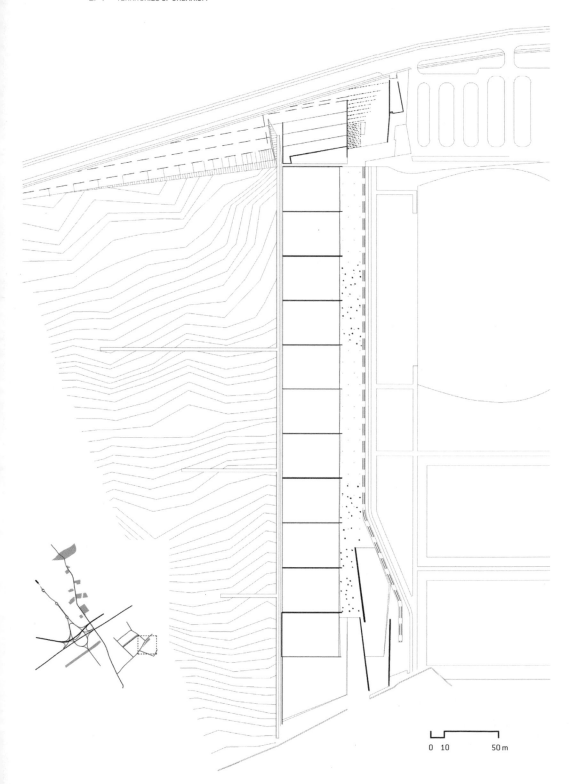

0   10   50 m

sturdy row of trees and shrubs similar to many others that border fields, almost a *bocage*. The material of the route is red beaten earth, shaded by the foliage of the linden trees below which it passes to reach the *plateaux*. A wall of exposed reinforced concrete, crossed by a horizontal line of light, marks the entrance and makes it possible to perceive the form of the sloping stone and grass plaza.

The second section counters the irregularity of the first with a regular rhythm of steps (the *plateaux*) that subdivide the level difference between the road at the top and the conclusion of the cemetery, in leaps of about 70 cm. The grass *plateaux* are bordered by exposed reinforced concrete walls and intersect the route, always establishing different relationships. To reach the level of the route, at least in one point, the *plateaux* are inclined to permit access; short slopes, again of grass, mediate the level shift between route and plateau. At times the grassy surfaces are at a level below that of the route, while in other cases they emerge, taking on volume.

The third section is composed of two elements: an incision and a relief. The first has been inserted to separate the cemetery from the rest of the still-cultivated countryside, the future of which could be that of a park of works of Land Art. The rolling Flemish landscape is marked by an incision in the ground that guides the gaze down into the valley, and that like the traditional "ha-ha" features of English parks divides without separating, without blocking the gaze. The relief is simply a soft bend in the terrain, but it is sufficient to conceal the entire cemetery from view for those arriving from the west, along the high road. The relief is a grassy path beside the incision; it makes it possible to descend and climb back up from the valley on the edge of the cemetery, and it becomes visible only when seen from inside the cemetery. A curve roughly reproduces the level difference between the two extremities of the cemetery, taking into account the insertion of a green plane that, like a long belvedere over the valley, accompanies the road on the crest (the *galette*). Again, the regular pace of descent of the *plateaux* reveals the shape of the terrain: the curve of the rise, like a large earthen animal, is higher in the first part of the cemetery and slowly lowers to allow the concrete walls of the *plateaux* to emerge in contrast.

The three sections and the whole cemetery confirm the descriptive capacity of the project to reveal a place, to produce new knowledge through a descriptive operation. Also, the cemetery is now visited and utilized by people uninterested in its function, simply there to enjoy the experience of the space.

Durable and perishable

In the abandoned waterfront port areas to the west of Rouen[5], an interesting location due to its proximity to the city center, an industrial strip is being transformed (one kilometer along the Seine, which crosses the city). The future of the zone seems to be that of gradual transformation for commercial uses and service industries. The area's character is determined by large lots (30,000 sq. meters) and the presence of "hard" activities, port offices, car dealerships, alongside a series of marginal activities that reutilize and subdivide the abandoned industrial sheds. Marcel

B. SECCHI, P. VIGANÒ, CEMETERY OF KORTRIJK: LOCATION AND PLAN.

Smets thus schematizes[6] the transformations of the area at the start of the 1990s: the original buildings, above all those along the main street, are empty and awaiting use for new purpose, as commercial showrooms, or are subdivided and marked by fast turnover; then a sort of cosmetic makeover is applied to the existing industrial building, and the leftover land is gradually filled up, to the point, in some cases, of replacing the existing buildings, occupying the entire depth of the lot to install economically important functions, creating the need for new areas for parking and spending time. The open space has little variation, as it is almost exclusively used for parking.

Recognizing this settlement logic, the project introduces certain criteria aimed at reducing the problems caused by the throng of scattered initiatives, the many under-utilized parking areas, and the monofunctional character of the space. The

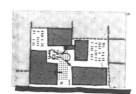

concept of "bays" makes it possible to improve the use of the large industrial lots without ulterior internal subdivisions. The project concentrates more buildings on the same lot, achieving a situation of complementary activities and utilization of public space; it frees up some of the area from parking (by means of rooftop lots) to organize a new public space; it envisions the formation of a village of businesses.

The project attempts to participate in – and at the same time to guide – evolution that is in progress. It sets certain rules for the different 'bays' that take the hard buildings and activities into account, in the sense of durable

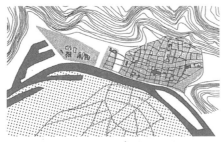

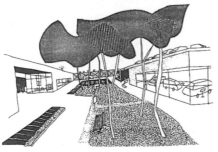
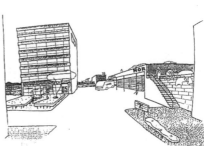

factors that are difficult to transform. It responds to the need to avoid modification of the borders of the lots and therefore of the form of the properties, while
organizing the *mixité*, from technical studios to small businesses, warehouses
to commerce, but also the specificity of each bay[7] and a program of economic
forecasting. The design is not unified, though it is crossed by themes that provide
some formal unity. The fragmented appearance takes into account the multiplicity
of the gathered elements, ordered in sequences[8].

A project in which the descriptive dimension is put at the center is not subdivided
into phases, implying development in stages, but instead into a certain number
of interventions that combine and can be independently sequenced. Each project
aims to achieve one of the identified themes and to get beyond the logic of the
interrupted project, breaking it down into groups of themes that assert lower possible levels of unification.

Plural histories

In the first lines of the report illustrating one of the projects submitted in the competition for the renovation of the Spedale di Santa Maria della Scala[9] in Siena, we
can notice the size of the work group required "for a theme of such density and
complexity of meanings and such vast contextual implications". Each discipline
is asked to "intertwine its own instrumentation [...] on the living, reacting terrain
of a great problem of urban science" for which the categories of restoration and
conservation appear to be insufficient. The descriptive project usually takes on
the character of a collective effort, whose construction involves the contributions
of different fields of knowledge and disciplines, each capable of deciphering logics and rationales that are often only apparently mute. Different gazes "detect"
and subtract from chance a wider number of issues, glimpsing their reasons and
absorbing them inside a project. Rogers and Chipperfield reject the idea of covering an incomprehensible complexity that has at times been used to legitimize
assertions of the impossibility of a description – or simply to legitimize superficial
positions. Again in this project, as in the territorial projects discussed previously, I
find "the profound attention to the 'great construction' in itself, to its long historical
development, its meanings and its role, its material structures that are incredibly
complex, but produced by motives whose logic must be understood".

The deepening of knowledge of the "great construction" constitutes the first of
the three areas of study around which the project is built. The second is reflection on the uses that "the edification suggests and permits, as a whole and in its
parts, starting from the comprehension of the genealogy of the present situation".
The third is an important assessment of the questions of physical condition and
maintenance[10]. The strategy of regeneration that forms the basis for the reflections on the Spedale proposes a process whose phases and gradual actions and
effects can only partially be foreseen a priori, a pursuit of "ever more advanced
balances"[11] through the identification of elements of authenticity. The "*passeggio*"

(promenade), the "*pellegrinaio*" (pilgrim's hall), the "*corticella*" (small court) and other types of spaces – fragments of an urban structure – are put alongside the distinction of durable and perishable elements on the part of Smets.

The recovery of the meaning of the various environments starts with an interpretation of the "bifrontal layout" of the complex, its character as a barrier towards the city and in the scale of the structures set onto the slope, with intermediate streets and ramps. The Spedale is equipped with an internal facade that remains limited to the scale of urban domestic space ("the serene and captivating dimension of the smooth wall") and an off-scale side composed of pilgrims' halls, regular longitudinal volumes parallel or perpendicular to the dramatic and monumental slope that goes beyond and incorporates the first circle of walls and the outer passages that are still functioning today. It presents itself with the 13th century *pellegrinaio* perpendicular to the facade, aligned along the axis of the Porticciola that would become the Passeggio, an imposing internal tunnel aimed towards the panorama

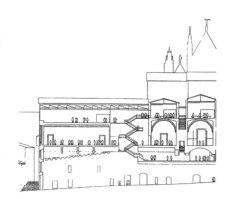

of the countryside, the fundamental structure of crosswise connection. In the 1400s, the experiment was to give the Spedale the character of "a constructed attempt, to which its contextual and plural nature and its very history were opposed', making it more than a building, "a great fragment of urban structure"[12], to be understood in its "plural and polygenetic reality: in the spaces, of course, but also in the materials, the workmanship, the surface effects"[13]. This is the only "whole" of the Spedale of which it is possible to speak and which it is possible to redesign. In a process of interpretation based on clarity and preci-

R. ROGERS, D. CHIPPERFIELD, F. IZZO (AMONG THE EXPERTS: M. MANIERI ELIA – HISTORY OF ARCHITECTURE AND RESTORATION; B. TOSCANO – HISTORY OF ART; A. CARANDINI – ARCHAEOLOGY; P. RICE, T. BARKER, OVE ARUP – STRUCTURES AND PHYSICAL PLANT), RENEWAL OF THE SPEDALE DI SANTA MARIA DELLA SCALA IN SIENA, INVITATIONAL COMPETITION, 1993.

SECTION OF THE INTERNAL STREET, ON PELLEGRINAIO AND PASSEGGIO (UP).

ORGANIZATION SCHEMES OF THE INTERIORS OF THE SPEDALE DI SANTA MARIA DELLA SCALA: PUBLIC CIRCULATION; VERTICAL LIGHT SOURCES; VERTICAL CIRCULATION; STRATEGIC INTERVENTIONS.

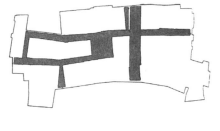
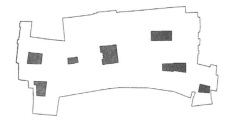

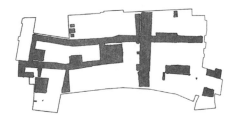

sion, the "clarification" of the building, and the elimination of its contradictions, as Odo Marquard would say, must be avoided at all costs (MARQUARD 1987).

The identification of the parts (constructed volumes and voids) makes it possible to approach the great construction, and is the basis of the project. The Pilgrim's Hall and the Promenade, the Church, the Court, the headquarters of the Compagnia dei Disciplinati, the Internal Street — each element is narrated in the report and the drawings. The focus is on micro-histories, biographies. "A correct revisitation of history is sufficient: of the material and institutional history, but above all of the history of meanings" to arrive at a project for the Spedale that does not pursue "improbable and abstract unities", but "truthful multiplicities and inconsistencies". Comprehension through a breakdown into formally and structurally defined blocks is not sufficient; it is instead necessary to also grasp the characteristics of "fluid continuity that end up surpassing those of stability and permanence"[14].

After having recognized that the Spedale is an aggregation of independent spaces, and not wishing to eliminate the boundaries and connections between the morphological blocks, the project introduces a system of horizontal and vertical public pathways whose visibility is facilitated by natural lighting. It is impossible to suitably light the basement zones, but it is possible "to make single rays of sunlight penetrate into the Internal Street, through the dark vaults of the basement zone" via special shafts that gain reflected light. Equipped nodes, staircase groupings, vertical conduits, and shafts are positioned in relation to the lighting points. Actions — and not phases — insert the design hypotheses into the long-term implementation.

Panorama

While the above cases underscore the ability to interpret and reveal the structural or representative aspects of a territory, there are many projects that pursue the staging of the territory as a panorama. Naturally there are numerous projects conceived with this goal, starting with certain magnificent panoramic streets and *corniches*, and extending to places of residence that open out like large windows onto the landscape. It is rare, however, for the plan of a city to systematically and stubbornly take the view, the panorama, as the basis of its design. One such plan, on the scale of a large city, is the one for San Francisco by Daniel Burnham, published in 1905, one month prior to the earthquake that destroyed the city, and quickly forgotten due to the urgency of the reconstruction. This plan has a descriptive approach that is quite extraordinary in several ways.

When he accepts the commission, Burnham requests that a bungalow be built on a spur of the Twin Peaks, and from this exceptional, panoramic, and privileged position, he begins his work[15]. The plan outlines a continuous and articulated structure of open spaces, without using the figure of the green wedge frequently found in 20th century planning, and instead using radical ways of connecting the set of spaces, without fear of the risk of losing the formal order pursued, on the

D. H. BURNHAM, "MAP OF THE CITY AND COUNTY OF SAN FRANCISCO SHOWING THE EXISTING HIGHWAYS, PARK AREAS AND PUBLIC PLACES", FROM THE REPORT OF 1905.

other hand, in major urban projects. The materials gathered in the Report are texts, photographs of landscapes, perspectives, and maps. The relationship among all these elements is generated by the focus on the forms of the territory in which the existing and future city is located.

After a short introduction that harkens back to the writings of Hénard and the Parisian radioconcentric model, the project is presented as a concrete device of description of the territory. The map shows the "system of circuit and radial arteries and its communication with San Mateo County" that partially recycle existing street segments; then come two photographic views, the first showing a "View of the natural harbor for yachts flanked by the proposed Boulevard, to the east of Fort Mason", the second "The bay, looking north from the proposed Outer Boulevard, near Fishermen's Wharf". These are immediately followed by the description of the Outer Boulevard, which illustrates the site and the project at the same time. Reading it, we learn that the Boulevard follows the

VIEW OF THE OCEAN, SOUTH, FROM THE PROPOSED TERRACE AT CLIFF HOUSE ROCKS

coast; near the wharves it has to be elevated to become a roof for the warehouses, forming a panoramic promenade for viewing the activities of the port. When the observer tires of the noises and work of the port, "he may note the changing aspects of the sea and study the effects of sunshine and shadow on islands and mountains seen through the masts of the ships"[16]. Earlier and successive photographs show the water and the ocean.

The entire route of the Outer Boulevard is thus constructed and narrated as a voyage of discovery of the physical and social characteristics of the city. The idea of *panorama* forms the basis for the design choices. The section along the ocean, for exam-

O. V. LANGE, VIEW OF THE OCEAN, LOOKING SOUTH, FROM THE PROPOSED TERRACE AT CLIFF HOUSE ROCKS, FROM THE REPORT OF D. H. BURNHAM OF 1905.

O. V. LANGE VIEW OF TWIN PEAKS FROM MARKET AND NOE STREETS (TO THE LEFT).

VIEW OF LAGUNA DE LA MERCED, LOOKING WEST FROM THE REPORT OF D. H. BURNHAM, 1905 (TO THE RIGHT).

VIEW OF LAGUNA DE LA MERCED, LOOKING WEST

VIEW OF TWIN PEAKS FROM MARKET AND NOE STREETS

O. V. LANGE, VIEW OF THE VALLEY,
TO THE WEST OF THE PROPOSED
SCHOOL FROM THE REPORT OF D.
H. BURNHAM, 1905.

VIEW OF THE VALLEY, WEST FROM THE PROPOSED ATHENÆUM

D. H. BURNHAM, DETAIL OF
THE PLAN OF SAN FRANCISCO
SHOWING THE SYSTEM OF
CIRCUITS AND RADIAL ARTERIES,
FROM THE REPORT OF 1905.

ple, based on the project of John McLaren, the great superintendent of Golden Gate Park from the end of the 1870s until his death in 1943, is positioned over the level of the sand dunes. While the references are the celebrated *corniches* of the "Riviera which skirt the Mediterranean from Nice, through Ville Franche and Monte Carlo, to Mentone and beyond"[17], what is striking is the central constructive role assigned to the description of the territory of San Francisco. The project of the plan and its structure do not fail to address atmospheric elements, colors, transparency, distances. Through the design of the boulevards and highways, which, not coincidentally, are the first parts to be illustrated, and without which the subsequent choices would not be comprehensible, the plan becomes a tool of discovery of the Pacific Ocean, the Laguna country, and the "superb natural scenery"[18] of San Francisco.

The hills are the focus of particular design efforts that have to do with the streets that surround them, ensuring easier grades, and calling for acquisition of the hills by the public administration to make parks and viewing points at their summits. The hills are a monumental complex around which the project should be structured, as in the case of the ocean. The photographs show the view from the hills towards the city and vice versa, from the various streets that offer views towards Twin Peaks, or from the surrounding hills. The fascination with the forms of surveyed land emerges – their almost physiognomic recognizability, the capacity to construct landmarks in the city and in a project that makes use of them. Thus the photographs, drawings, and texts describe the hills almost one by one, together with the valleys that separate them. The streets follow their contours at the different levels; the terracings interpret their slopes. From each hill the view of the others, and not just of the city, must be permitted. Strangely enough Olmsted is not mentioned, and the neoclassical style of the specific projects, with references to Le Nôtre, does not appear very attractive today. But beyond the style of the urban design and the works of architecture, what remains timely and stimulating is the description of the territory starting from its design, the care put into the design of open space placed at the center of the plan and imagined inside a time span of fifty years.

It is only in the final section, that of the acknowledgments, that less predictable yet decisive references and presences come to the fore. Burnham thanks John McLaren, whose efforts were already, at the start of the 1900s, a "life work". Many of his suggestions, Burnham writes, have been incorporated in the plan (the acquisition of Glen Canyon, and regarding the Great Highway). Above all, the figure of Oscar Victor Lange stands out, a leading photographer of the San Francisco Bay Area at the time, who specialized in architecture and landscape photography and created the beautiful images that accompany and seem to generate the project. It would be interesting to further explore the relationship between the urban planner and the photographer, through the concrete modes of their work. They are both "inventors of landscapes" (Viganò 2008d). In this plan, the conceptual part that is connected to the concepts of the "greenway" and the "parkway", the type of project, and the mode of its construction represent the least original of Burnham's contributions. Instead, the relations between forms of the territory, landscape, their photographic representation, and the project indicate intriguing lines of research. The landscape is not just panorama. The photographs represent its perceptible characteristics, opening towards an interpretation of the project as their description.

Meta-tourists

The urbanist and the architect, like the scholar of landscapes (JACKSON 1970), are very particular and privileged tourists. Their job usually starts when they visit a place unfamiliar to them, or that they do not fully know, walking through it and observing with all the senses (ZARDINI 2005), combining analytical knowledge with a parallel path of experience. Recent years have produced much research that begins with the idea of a movement in the territory and its description[19] and shares the focus on the culture of each portion crossed. The project can set the conditions for its appropriation and integration, contributing to the relational and situated construction of knowledge (TIETJEN 2009).

Between context and autonomy, which can generate two extreme attitudes, as well as mimetism and self-referentiality, certain families of descriptive approaches can be introduced that are borrowed freely from a study by Anna Ottani Cavina on the neo-classical city. Here, "nature" stands for "context" (OTTANI CAVINA 1994).

Projects *ajustés sur la nature:* "…the view has undergone a process of erasure and synthesis, which an 'adjustment' has taken place, an 'arrangement' of nature"[20]. Certain projects of Georges Descombes belong to this family, or the exercises of reading of other projects, such as those never implemented or simply sketched out by the masters of the Modern Movement, as in the didactic exercises of Francesco Venezia. Many of the projects of the Portuguese school in the 1980s, and some of the sketches for the project for the University of Calabria by Franco Purini, also belong to this family. This process of adjustment pays attention to all the signs, including erasures and differences. Anomalies and discontinuities are not denied but rather are pursued as clues, traces of something that has been lost, but which it might be important to find again. This same idea of not following the main line of a phenomenon migrates through the disciplines, whether it is that of evolution, as in the case of the neo-evolutionist observations of Stephen Jay Gould (GOULD 1980), or that of the study of the characteristics of other societies, as in Clifford Geertz (GEERTZ 1973)[21], or the clues of Carlo Ginzburg (GINZBURG 1992). In the background: "The artist (architect/urbanist) as translator of the chaos that wraps the surface of things, the artist as demiurge that makes the order of nature, its rational and eternal nucleus, emerge. The artist as he who reveals a '*sous-jacente*' structure, not the exterior of the form". In these definitions we can read all the ambiguities of the descriptive project: the descriptive project as structuralist-idealist project.

Projects *composés sur nature*: I attribute to this definition the meaning to which Pierre-Henri de Valenciennes referred in his notebook, contrasting *composé* and

ajusté, i.e. "taken faithfully from reality as in the landscape with a lake, the mountains, trees…" (Оттанɪ Cаviна 1994: 98). The verb *compose* implies the possibility
of manipulating the materials named, found, and reutilized in different contexts.
This family includes certain projects by Rem Koolhaas, Jan Neutelings, Bernard
Tschumi. A choice is made of *les composantes* that are most significant in the contemporary landscape, and they are reutilized with just slightly different composi-
tions inside new projects. In both Tschumi, in the project for Parc de la Villette, and
Koolhaas, for example in the Kunsthal of Rotterdam, this process of taking apart
and putting back together crosses the scales to the point of involving the choice
of materials, drawn from what we normally see and touch. Transparent corrugated
material clad the walls of a museum, plywood furnishes the main hall, the tubes of
luminous billboards light a new park. These are not so much humble as normal and
modest, ordinary and, in a certain sense, banal materials, faithfully lifted from reality and re-composed *as if* the contemporary context had been formulated through
them, not in the sense of imitation but in that of representation and awareness. In
the background lies the possible autonomy of the single element, the idea of heterogeneity as value (a position already sustained by Laugier) and therefore the
possibilities offered by deconstruction. Other very well known projects, such as the
competition project for the *ville nouvelle* of Melun Sénart in 1987 by Rem Koolhaas,
have a marked representative character, rather than being descriptive of the physical heterogeneity and the programs, the discontinuity and paratactic juxtaposition
of objects and practices. The need for a new interpretation of contemporary situations forms the backdrop for the representative dimension of many projects.

At the start of the 1990s, the Randstad was described by Jan Neutelings as "an
extensive patchwork carpet", in which each patch was equipped with a specific
program and physical structure[22]. The representative dimension of the project,
legible in the diagrams, takes form from an initial conjecture from a model, which,
in the case of the *Patchwork Metropolis*, does not oppose what exists, but emphasizes it – in any case representing and describing it. The model of the compact city
is countered by low-density living in contact with metropolitan structures and services; the metaphor of the patchwork is a "design description", as was Samonà's
image of the urbanized countryside, a reconceptualization of what exists, that
– apart from speaking of the future – also describes much of the present. The
fragments of this new urban landscape belong to a non-physical and complex
order that makes greater articulation of experiences and ways of life possible, in a
space endowed with many qualities. It is "a transformation model; the model of a
field in permanent evolution".

Less important regarding the theme approached here are the "portrait-projects"
and the projects *d'après nature*. In the first case, the project "wants to be the mir-
ror of an experienced place, a real path of the painter (architect, urbanist). He too
comes in the wake of the *Rêveries* of Rousseau, like so many *promeneurs soli-*

taires", which had introduced "that sentiment of places that serves to narrate a bit of ourselves" (Оттаnі Cavina 1994: 109-110). In the background are the project-autobiography and the project-diary. In the second case, that of projects *d'après nature*, we are looking at imitative projects that share a mimetic, acritical stance regarding reality and history. I am thinking of the school of Brussels represented by Culot and Krier, and, above all, of projects that are simply done without introducing any shift, not even that outlined by an *as if*.

As a whole, the four families take the descriptive dimension of the project as a privileged vantage point, though in profoundly different ways. The first two approaches seem useful to understand the city and the territory, and starting with the short references above I would like to try to reach some more general considerations. The project as description guides the survey and is a survey in its own right, the tool of description and technique of representation and interpretation, a "form of safeguarding of each individuality" connected to the awareness of our finitude: "in any future generated by a transformation, a quantity of past is conserved that always goes far beyond the measure of the transformation" (Marquard 1987: 66). This makes an accurate operation of surveying places and the society that inhabits them, of reading and interpreting the traces as indispensable symptoms and clues. It requires putting objects and themes belonging to different disciplinary spheres into relation in ways that differ from those of the interdisciplinarity of the 1960s, but with the aim of designing with associations and unveilings.

In the observation of a place, of a territory, the hermeneutic position of Odo Marquard provides a guide to possible approaches: understanding is more important than deduction; no one can start from scratch; if necessary, it is better to put up with contradictions rather than accept the impression of their solution (Marquard 1987: 21); and with respect to tradition, "the primary suspicion of its reasonableness and the duty of an explicit justification lies with those who reject it" (Marquard 1987: 87). This short and modest manual has profound consequences on the modes of the gaze, and effaces none of the radical nature of the project. It obliges reading of the plurality of the histories ordered and juxtaposed in time and space. The microhistories constructed up against each situation allow us to take stock of recognizable individualities. Through the narrative, they structure the description: "description, simple concentrated description" that, according to Gertrude Stein, made English literature great (Stein 1935).

---

¹ The theory of modification is discussed in the editorials of Gregotti and the articles by Secchi in the magazine *Casabella*, in the 1980s and 1990s.

² The latter three attitudes, referring to three images of the discipline, were discussed in *Casabella* by Secchi 1992a; 1992b.

³ The form seen as a diagram of forces is a concept expressed in *Notes on the synthesis of form*, 1971, which Christopher Alexander picks up from D'Arcy Wentworth Thompson.

⁴ The project that won the competition was prepared by B. Secchi, G. Serrini, P. Viganò, C. Zagaglia in 1989.

⁵ The project of restructuring of the industrial friche of the districts to the west of Rouen was prepared by *Projectteam Stadsontwerp* (M. Smets, K. Borret, E. Van Daele, H. Van Bever). Ville de Rouen, *Aménagement des Quartiers Ouest, final report of the study*, December 1993.

⁶ Also see BORRET 1994.

⁷ "The principle of the bays is already inscribed in the place; to the north of Avenue du Mont Riboudet ... which defines a network of spaces 'en creux', between the dense constructions. The proposed conception has the aim of sustaining this organization" (SMETS 1993).

⁸ "In general one thus obtains a form that at first glance might seem irregular, but which is constituted on a very continuous and homogeneous pattern in keeping with the two main directions that generate contiguous and articulated spaces from the form and the orientation of the walls"(SMETS 1993).

⁹ The invitational competition was held in 1993 and won by Guido Canali. The project discussed here is instead the one prepared by R. Rogers, D. Chipperfield, F. Izzo (experts: M. Manieri Elia – History of Architecture and Restoration; B. Toscano – History of Art; A. Carandini – Archaeology; P. Rice, T. Barker (Ove Arup) –structures and physical plant; S. Pietrogrande (Dioguardi group) – maintenance; L. P. Scandizzo – Socio-Economics; C. Malby (Davis Langdon and Everest) – construction costs.

¹⁰ "In our visits to the Spedale we have been struck by the stability of the environmental conditions of the building. The internal environment has been protected by its mass and depth. The small exterior surface, as compared to the internal volume, has helped to stabilize the environmental conditions [...]. Regarding the structure of the building, like many other antique buildings it has reached a sort of state of repose. We believe that this balance is one of the great resources of the building, a resource that imposes its own problems and its own solution criteria. Obviously the changes in the environmental conditions have to be evaluated in relation to this stability" (from the project report, pp. 101).

¹¹ From the project report, pp. 14.

¹² From the project report, pp. 19.

¹³ From the project report, pp. 24.

¹⁴ From the project report, pp. 23.

¹⁵ "Report of D.H. Burnham on the Improvement and Adornment of San Francisco", 1905, pp. 8.

¹⁶ "Report of D.H. Burnham on the Improvement and Adornment of San Francisco", 1905, pp. 53.

¹⁷ *Ivi*, pp. 55.

¹⁸ *Ivi*, pp. 56.

¹⁹ For example: CORNER, MACLEAN 1996.

²⁰ OTTANI CAVINA 1994: 98. "'Ajusté sur la nature', the handwritten phrase of Valenciennes repeated sixteen times in his notebook Rome, is never related to the views literally taken from reality. It is placed instead beside the rectified and cleaned drawings, where the view has undergone a process of erasure [...]".

²¹ "Precisely these terms – 'strangenesses' and 'oddities' – constantly return in Geertz's definitions of the object of anthropology", Remotti, F. 1987, *Clifford Geerzt: i significati delle stranezze* (introduction to the Italian edition of Geerzt C. 1973), *The Interpretation of Cultures*. Geerzt is associated with the concept of "thick description": "[...] ethnography is thick description" (GEERZT 1987: 46). Any ethnographic description is "extraordinarily thick". The meaning of this expression, among others, is a warning against a merely phenomenic description, and the term is used as opposed to "thin" description.

²² For the illustration of the study conducted by Willem Jan Neutelings with P. Sulster, P. van Wesemael, and E. Winkler in 1989-1990, I make use of the text by W. J. Neutelings, revised and extended by Piet Kalsbeek (not published).

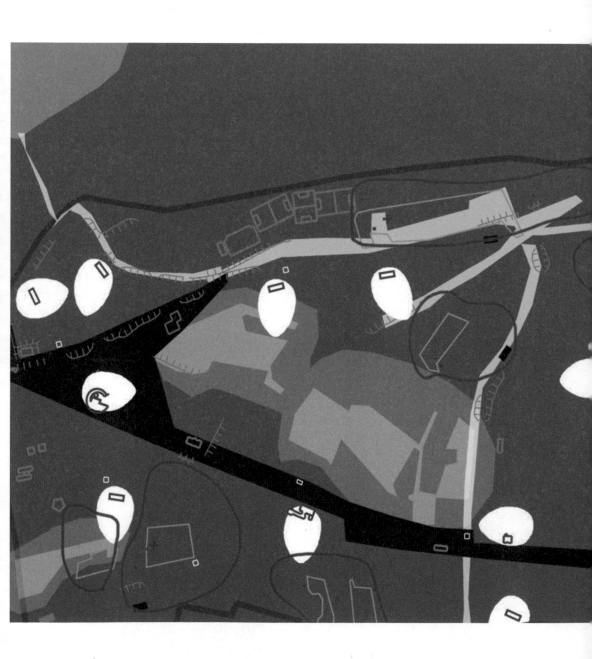

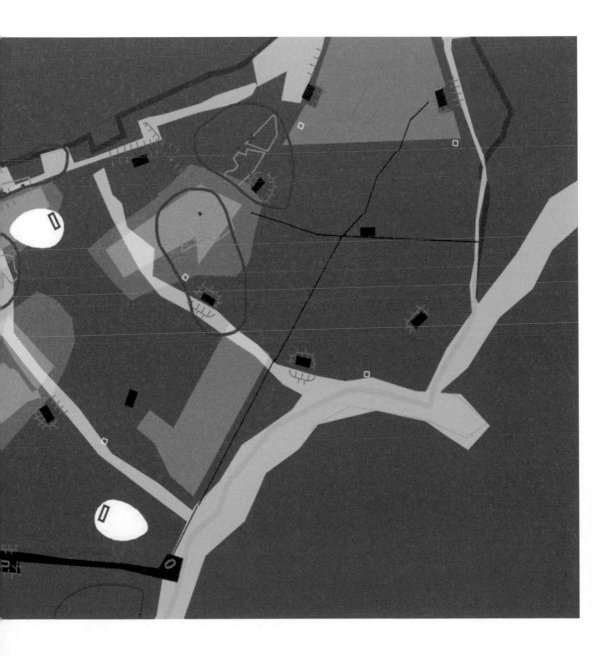

Behind every project lies a specific interpretation and conceptualization of the territory. The Master Plan for the education center of De Hoge Rielen recognizes, distinguishes, and combines the three fundamental landscapes of which it is composed: the first is the natural landscape; the second is connected with the military past of the area; the third, finally, is the educational landscape, formed starting with more recent uses.

The three landscapes mark the history of Hoge Rielen. They are three partially different layers that emerge and vanish in the territory, overlapping, cooperating, or clashing. The various projects that make up the Master Plan of De Hoge Rielen are radically descriptive and take an elementary approach.

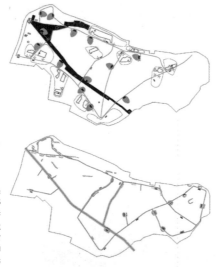

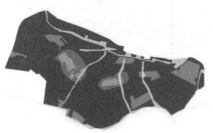

MASTER PLAN FOR DE HOGE RIELEN:
THREE LANDSCAPES
FROM BELOW:

THE NATURAL LANDSCAPE (IN DARK GRAY, THE MATRIX OF PINES; IN LIGHT GRAY, THE AREAS OF PERCOLATION OF THE NEW BROADLEAF FOREST);

THE MILITARY LANDSCAPE (IN OCHRE, THE MILITARY ROADS; IN ORANGE, THE BUILDINGS AND EMBANKMENTS);

THE EDUCATION LANDSCAPE (IN DARK GRAY, THE CONTACT STRIP; IN LIGHT GRAY, THE SPACES OF PROXIMITY; IN GRAY, THE MILITARY BUILDINGS ALREADY TRANSFORMED INTO FERMETTES; THE RED BORDER INDICATES SPACES OF PROXIMITY OF THE CAMPGROUNDS.

B. SECCHI, P. VIGANÒ (2002, IN PROGRESS), MASTER PLAN AND PROJECT FOR DE HOGE RIELEN (BELGIUM), WITH U. DUFOUR, L. FABIAN, J. LEENKNEGT, G. ZACCARIOTTO (COMPETITION: 2002); U. DEGLI UBERTI, U. DUFOUR, T. LOMBARDO, P. OCHELEN (MASTER PLAN: 2004); U. DEGLI UBERTI, T. FAIT, S. GEERAERT, E. GIANNOTTI, G. PUSCH, S. PELUSO (CONTACT STRIP: 2007, IN PROGRESS).

The natural landscape, at first glance, seems easy to decipher. The signs of water and topography create slight dune shapes that cross the pine forest, of great aesthetic value. Nevertheless, the ecological value lies instead in the still young deciduous forest, composed of heterogeneous fragments, patches marked by a high level of biodiversity. This forest will become the new matrix on which to graft other species.

The second landscape is formed by the remnants of the military past. These are minimal works of architecture that generate a severe, not immediately legible atmosphere. Powder magazines, embankments, and reservoirs are scattered across the hundreds of hectares of De Hoge Rielen in keeping with the rule of the 'proper distance' and camouflaged in the vegetation and the shaping of the terrain. The roads and their relations with the topography, made of cuts and hollows, represent the most important legacy of the military past of Hoge Rielen.

Finally, the third landscape, which is connected with recreational and educational use, utilizes the two previous landscapes as a context for the new functions and to welcome practices different from those of the past, defining spaces of proximity.

DE HOGE RIELEN MASTER PLAN.

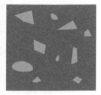

MATRIX INVERSION SCHEME:
CONIFER FOREST *VERSUS*
BROADLEAF FOREST.

An "elementarist" description

Each landscape is defined by a pattern of a few elementary units that produce recognizable spatial sequences. For example, the education landscape is composed of spaces of proximity around the reutilized military buildings, or the new constructions, permitting independent activities of small groups: from the large lawns for fire, play and camping, to the equipment scattered in the woods. The spatial sequences relate to the tradition between indoor and outdoor, corresponding to paths that begin in indoor space and head towards the forest, or from De Hoge Rielen and move towards the urban centers outside it.

The Master Plan proposes a series of measures to make De Hoge Rielen an inhabitable space in a sensitive territory of growing biodiversity. The strategy implies the design of a collection of basic devices that permit incremental actions to ensure the connection of the various brush areas and reinforcement of the natural landscape; the conservation of the military landscape and the recovery of the pavilions; a clearer distinction of the space of proximity conceived as an "interior" in the forest and a space of contact with the outside world, along the existing axis between the two entrances.

Natural landscape

The presence of tame, usable nature is the main reason behind the existence of De Hoge Rielen as a place for education and recreation. The long tradition of the *bosklas* (classes in the woods) and outdoor sporting activities, which are part of the education program in many northern European countries, is combined with a social and political objective, in the wider sense. De Hoge Rielen is a place of encounters, discussion, education and exploration, with an ethical focus in which the natural element plays a precise role.

The current ecological matrix of De Hoge Rielen is a large forest of conifers of relatively recent formation that is part of a vast homogeneous area crossed by a few waterways and studded by several wetland areas. The project of replacement of the pines with woods of broadleaf trees, which are more suitable for the context

FORESTS AND WATERS.

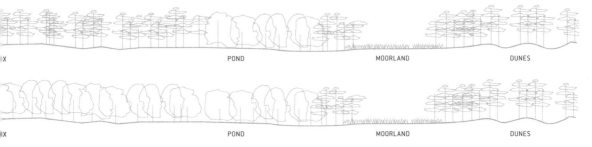

IX | POND | MOORLAND | DUNES

IX | POND | MOORLAND | DUNES

and richer in biodiversity, begins with the percolation of naturalness along the ditches cut by the military along the roads, which thus take on a new and important ecological role. In the decades to come, the broadleaf matrix will absorb the remnants of the pine forest in the areas where the soil is less rich and sandier, which readies itself to become a new cultural landscape.

MATRIX INVERSION SECTION.

Military landscape

At first glance, the military past of De Hoge Rielen has left only weak signs on the territory. They involve a minimum expression and are thus easy to erase . Originally enclosed by embankments that concealed and protected from explosions, and which are now invaded by native vegetation, the buildings are close to the ponds for protection against fire and are connected by roads, made with reinforced concrete, that form a grid. The rigor of the military architecture and planning is visible today above all in the eastern part of the area; the buildings are inexpensive and dull.

The Master Plan suggests three fundamental moves to bring higher levels of comfort in the buildings; they do not, however, correspond to more invasive levels of intervention. The idea is that every space should correspond to different degrees of comfort, which in turn permit the insertion of different practices.

The degree zero of comfort implies only maintenance of the building to avoid decay: the pavilion is a covered and bordered space that contains services connected with camping and daytime activities. A second level of comfort requires salvaged zones

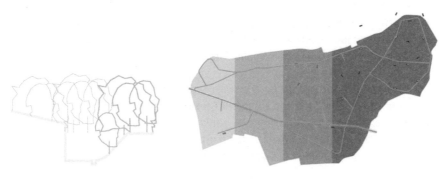

MILITARY LANDSCAPE GRADIENT.

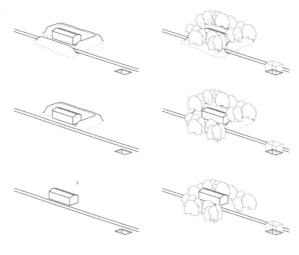

that permit prolonged use. Finally, the third level of comfort would involve the construction of a new skin. Instead of this possibility, we prefer the insertion of a modular box that is thermally insulated and outfitted.

Educational landscape

The basic unit of the educational landscape is the space of proximity. The building, usually a military pavilion converted as a *fermette* (diminutive of *ferme*, meaning farm, indicating the generalization of the rural architecture of the houses of Flemish dispersed settlement) is at its center. Often the military roads cross

UNITS OF THE MILITARY LANDSCAPE: THE POWDER MAGAZINE, THE PROTECTION TALUS, THE VEGETATION.

it, and for this reason the Master Plan proposes connecting the various units by means of a network of routes that permits reversal of the opening of the pavilions towards the south and the forest, without interruptions. In this way the space of proximity can be minimally identified and designed. The space of contact along which the main collective facilities are located, the access building and the small theater, etc., becomes an 'exterior' in which the various groups meet.

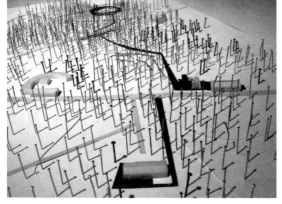

An architecture-landscape

The "contact strip" (east-west orientation) is perpendicularly crossed by a route along which are placed the multifunctional space, the open space of the theater and the picnic area, and the new hostel. These are four small

DE HOGE RIELEN: FOUR PROJECTS AND ONE ROUTE.

WORKING MODEL OF THE NORTH-SOUTH ROUTE THAT CROSSES THE CONTACT STRIP WITH THE FOUR PROJECTS (FROM BELOW: THE BUILDING TO RENOVATE AS A FUNCTIONAL SPACE, THE OUTDOOR SPACE OF THE THEATER, THE PICNIC AREA, THE HOSTEL).

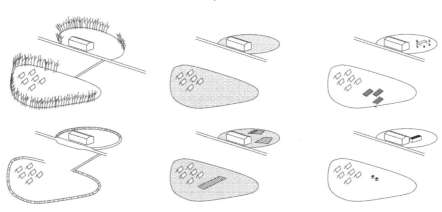

DIAGRAMS OF SPACES OF PROXIMITY.

projects that give substance to the hypothesis of "contact space" and character-
ize different portions of De Hoge Rielen.

The north-south route is one of the thresholds that mark the gradual passage from
the main entrance and from the active educational landscape to the quieter military landscape toward the east.

The outdoor space that connects the small interventions is a forest in which a long
seat-path is inserted, a platform that rests on the undulated ground, followed by
different inclined planes that approximately replicate its form.

The new hostel concludes the route to the north and takes its circular layout – containing the bedrooms and the indoor, outdoor, and sheltered common spaces – from its
close relationship to the pond. The form reveals the profile of the water and becomes

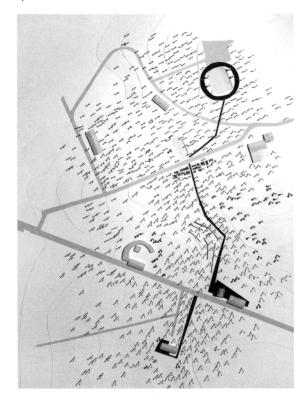

DE HOGE RIELEN: FOUR PROJECTS
AND ONE ROUTE.
WORKING MODEL OF THE NORTH-
SOUTH ROUTE THAT CROSSES
THE CONTACT STRIP WITH THE
FOUR PROJECTS (FROM BELOW:
THE BUILDING TO RENOVATE AS A
FUNCTIONAL SPACE, THE OUTDOOR
SPACE OF THE THEATER, THE
PICNIC AREA, THE HOSTEL).

VIEW OF THE ROUTE THAT MEETS
THE CONTACT STRIP ACROSS THE
FOREST.

its staging. The building forms a unit with
the landscape; it is an architecture-landscape.
Arriving from the south, a roof supported by a
light wooden structure and surrounded by a
pine forest in slow transformation marks the
passage to an inner space in which the broadleaf trees signal the presence of water. From
the north, a closed wall is interrupted at a few
points: the space of the portico accessible to
all; inside, the circular winter garden around
the water that creates a shared, more intimate space; and finally the space of privacy,
the bedrooms.

The insertion of a hostel in the three landscapes has led to reflection on different conceptions of living together, coming to terms
with the tradition of De Hoge Rielen. The idea
was not just to describe a place, but also to
represent, through the spaces, an approach
and a vision of the relations between groups
and individuals. How does the hostel in De
Hoge Rielen differ from any other hotel in a
panoramic location?

POND.

The pond could be interpreted as a place of individual enjoyment, for example from a bedroom with a loggia; any other presence could be seen as disturbing the appreciation of the landscape, within a conception of maximum privacy of the individual-nature relationship. The project is guided by the exploration of other relations. The appreciation of the panorama happens on collective ground, and is total only when one is in the winter garden, a space of appropriation wider than a corridor and always variable, a window larger than any other, a space of relations and inside-outside continuity.

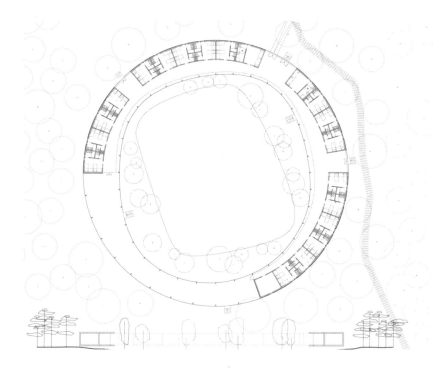

DE HOGE RIELEN: PLAN OF THE
HOSTEL, SECTION AND VIEWS.

This approach may seem ideological, but it is the very institution of De Hoge Rielen that is ideological, when it grants a forest an educational purpose. There is always a presence in the project ideology. It is a system of ideas regarding space that connects sensations and perceptions to the intellect.

The project describes the different dimensions of Hoge Rielen, its physical landscape, and the idea of relations between group and individual sustained by the educational program of this place.

DE HOGE RIELEN: VIEW FROM THE INTERIOR OF THE HOSTEL TOWARDS THE BASIN.

TERRITORIES OF THE FUTURE
Part III

> Futures forecasting is only useful when its con-
> tent is in a form and detail that enables it to form
> a basis for constructive dissent.
>
> PRICE 1977

Beyond the stage of vacant staring[1]

The activity of a researcher, of a scholar in different fields and disciplines, is always marked by the formulation of hypotheses: both when the project is an opportunity for conceptual innovation and a tool of description, as discussed in the first and second parts of this book, and when it is a collection of hypotheses about the future, as in the reflections contained in its third and last part[2]. The project is a specific form of knowledge that is often misunderstood among different types of knowledge. As empirical knowledge, design activity proceeds by means of experiments from which to extract fragments of generalizations and, eventually, of theories.

Delving into the logical operations that set design activity apart, one notes a structure that is limitedly deductive while being, to a great extent, inductive and abductive. Reversing the terms proposed in the 1930s by Ronald Fischer who spoke of the "design of experiments", the term "experimental design" (FISCHER 1935) is instead understood here as experimentation conducted with the tools of design: from this, as from a correctly conducted experiment, it is possible to extract valid inferences[3]. Of course, like a scientific experiment, the project can also be constructed in non-rigorous ways.

To design, to recognize, to describe a situation, to formulate the interpretation of a phenomenon — all imply the use and choice of various constructions of reasoning. Different degrees of uncertainty and faith in our capacity of observation permit operations of *deduction* ("which depends on the faith we have in our ability to analyze the meaning of the signs in which or through which we think"), *induction* ("which depends on the faith we have that the course of a type of experience will not be changed or will not cease without some previous indication of its ceasing") and of *abduction* ("which depends on our hope to guess, before or after, the conditions under which a given type of phenomenon will present itself")[4]. Peirce believed that "the whole fabric of our knowledge is one matted felt of pure hypothesis confirmed and refined by induction. Not the smallest advance can be made in knowledge beyond the stage of vacant staring, without making an abduction at every step"[5].

The scientific use of imagination

Among the various design operations, the construction of scenarios is perhaps the moment in which, with greatest clarity, the project presents itself as a coherent sequence of hypotheses. In the definition proposed here, the scenario is a collection of hypotheses that investigates the future and allows us to address and discuss the future. Sherlock Holmes does not guess; he makes scientific use of the imagination, conducting operations of abduction that always start from material bases, tracks, clues, signs (GINZBURG 1992). Every hypothesis in the region of the probable, the possible, and the plausible[6] is broken down by Sherlock Holmes into its logical parts; it springs from a question whose answer represents, if not the solution of the case, at least an advance in the knowledge of the observed phenomenon, or the exclusion of one possibility.

I will not delve further into the reconstruction of the history of the scenario, a design technique and operation that has had many different versions over time, and through which it is possible to read the passage from a deductive and inductive idea, the projection of the past and present into the future, to a prospective and abductive idea. Nor will I explore the latitude of its use: from the scenarios constructed by experts in military strategy, economists, political scientists or demographers, to those produced by ecologists and scholars of the natural sciences that simulate the different vegetable and animal configuration in time and in their evolution, to the scenarios of the experts of marketing or of urban and territorial development. I will not focus on the users of scenarios, which can be widely varied, from businesses that attempt to forecast behaviors and preferences, combining the construction of scenarios with market research, to politicians, administrators and inhabitants who prepare a project for the territory and think about its possibilities and potentialities. Nor will I address the roles of the various actors, apart from certain experiences. Instead, I want to draw forth, from the copious and often technical literature on the scenario, the deeper philosophical sense of questioning about the future. Secondly, I would like to make some hypotheses regarding the reasons why, in recent years, there has been a return to such frequent use of the construction of scenarios. Finally, I will use certain experiences in the field of design (of architecture) of the territory and the city to think about the form of knowledge the scenario produces (*Territory 5, Scenarios of Dispersion*; *Territory 6, Paleochannels of History*; *Territory 7, Scenarios of Living Together*). The intentions outlined here are not arranged in linear ways in the following text, but surface from time to time to orient the reading.

[1] The phrase is by Charles S. Peirce, 8.384-388. Quoted in: Eco, Sebeok 1983: 17-18.

[2] I have addressed this theme (Viganò 2008c) in the preface to the book by Bozzuto, Costa, Fabian, Pellegrini 2008, *Storie del futuro*, revised and extended here.

[3] "I have assumed, as the experimenter always does assume, that it is possible to draw valid inferences from the results of experimentation […]", Fischer 1935, from the edition 1960: 3.

[4] Charles S. Peirce 8.384-388. Quoted in: Eco, Sebeok 1983: 17-18. Also see Rowe 1987: 101.

[5] Charles S. Peirce Ms 692. Cited in: Eco, Sebeok 1983: 33.

[6] "Into the region where we balance probabilities and we choose the most likely", Doyle 1902, from the edition 1981: 38.

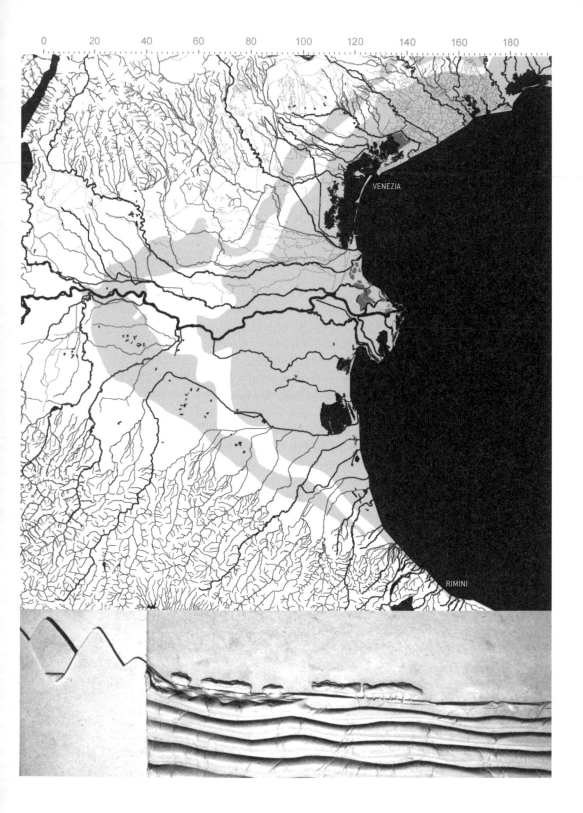

The future is timely again: after years of insistence on the present, on the contemporary condition as the foundation of any reflection and any project, the
future forcefully emerges again and imposes its own yardstick of judgment on
the present. My hypothesis is that in contemporary society there is a very strong
relationship between uncertainty, the perception of risk and fear, and the use of
scenario construction. Awareness of environmental issues, of their gravity and
potential to cause conflict, has expanded discussions of scenarios about the
future of the planet to a worldwide stage. But there also exists — and this is a
hypothesis — a specific condition that nurtures the production of scenarios on
the part of architects and urbanists. There is a close relationship between awareness of the conclusion of the experience of the modern city, the emerging of new
settlement forms, the need to analyze their consequences in the future and to
project possible alternative scenarios. I also believe that the use of the scenario
is the result of greater awareness, based on experience, of the illegitimacy of
many deductions taken for granted in the past, and an increasingly refined epistemological critique of those conclusions. The frequency of this operation also
justifiably leads to observation of the scenario as a form of discourse on contemporary territories.

Uncertainty, risk, fear

Fears are definitely the motivation behind the increasingly numerous efforts to
produce images of the future, together with the concept of "world problematique"
that is the reason for being at the origin of the Club of Rome. The idea of interde-
pendency of phenomena on a planetary scale analyzed by Jay Forrester in systems theory and then translated into the models World 1 and 2 used in the report
The Limits to Growth (Meadows, Meadows, Randers, Behrens III, 1972) is connected to the debate on growth and the need to think about the future. The models
and scenarios developed starting in the mid-1960s have a strong cosmological
character, instilling a worldview and taking a position with respect to the theme of
infinite growth and imbalances (VIEILLE BLANCHARD 2007).
While being quite critical of the most famous centers of futurology in the United
States, such as the Rand Corporation, as well as of the commission composed
of academics and intellectuals created by the Academy of Arts and Sciences to

MAP OF HYDROGRAPHY ALONG
THE COAST BETWEEN RIMINI
AND VENICE. THE LOWER AND
MORE VULNERABLE AREAS IN
CASE OF A RISE IN SEA LEVEL:
IN GRAY, AREAS BELOW THE
LEVEL OF + 5 M.

SECTIONS OF SOIL TYPES.

work on scenarios for the year 2000[2] which formulated the scenario of the post-industrial society[3], the research – of which *The Limits to Growth* is the best known example – challenges, above all, the basic hypotheses of continuous growth, stable geopolitical order, and unlimited resources. The contrast in the United States generates a very interesting debate between those who foresee a dark, hopeless future, the proponents of "doom and gloom", demanding a radical change of direction, and those who maintain a positive attitude towards the characteristics of human progress (of the capitalist variety) and its capacity to reinterpret the given conditions. The first group contains documents like *The Limits to Growth* and intellectuals like Lester Brown and Barry Commoner, while the second contains *The Year 2000* (KAHN, WIENER 1967) and figures like Herman Kahn and Julian Simon, who will return more than once during the course of this chapter. Prophecies and scenarios are two different ways of thinking about research on the future, as Aligica emphasizes (2007) in his text on Kahn and Simon, blazing a particularly cogent trail of research, since the interpretations of the future of the two opposing factions reveal a historical moment driven by the counterculture and the reactions it engenders. In future studies the different positions emerge, clear and distilled, and it remains to be demonstrated that *The Limits to Growth* can be reduced to and ridiculed as prophecy. While this text gives rise to a new tradition, Kahn and Simon trigger a "critical counter-tradition" (ALIGICA 2007: 3), reinforced by intense publishing activity in which *The Resourceful Earth* (1984) is the response to *Global 2000, Report to the President* (1980), prepared by the Council on Environmental Quality and the Department of State for Jimmy Carter. Even today, thirty years later, the construction of scenarios contributes to the field of expression and debate of contrasting positions; it acts anew and in active ways in the reconfiguration of the basic paradigms that lie behind the expectations for the future and inevitably influence our behaviors in the present.

The scenario has the virtue of reinserting time in the study of transformations which the science of economics, in the last years of the 1800s, had attempted to link back to the relationship between cycles and circumstances – a position that was definitively challenged by the Great Depression in America in the 1930s. The pursuit of short-term balances, also supported by Keynes, who is credited with the famous remark "in the long run we are all dead"[4], gives way between the 1930s and the 1950s to the rise of a different way of forecasting the future[5] beyond the very short term. These are attempts at planning and medium-term forecasting that develop in different forms, from the extreme Soviet approaches to the more temperate operations of the Dutch and the French, mediating between the free market and government control. The use of scenarios emerges only later, alongside the first experiences of national planning, in the French case. Generated in a context of hardship (the first Monnet plan covers the period 1948-1953), this planning finds itself immersed, after a few years, in the new society of consumption (ROSS 1996) that reshuffles social values and expectations. In a rapidly changing context, the selection of a "figure of the future"[6] seems much less clear and legitimate, along with the choice of the objectives to be pursued. The construction of scenarios and the *prospective*[7] thus also belong to the great effort of reconstruction and modernization after World War II. They are the expression of a desire to construct the future, but they also shed light on the condition of uncertainty in which decisions are made, with few exceptions limited in time and space.

Uncertainty.   Addressing a condition of uncertainty has always been crucial for design, and in particular for the design of the city and the territory. The scheme constructed by Karen Christensen (CHRISTENSEN 1985) is well known: according to the author, the discussion and critique involving the paradigm of the rational decision of the planner (based on the idea that it is possible to be "taken as given consensus on objectives and knowledge of the means to achieve them' [BALDUCCI 1991: 125]) should evolve in the direction of a more detailed schema through which to analytically interpret the conditions in which the project is constructed, i.e. the different degree of uncertainty in which objectives, resources, and technologies are positioned.

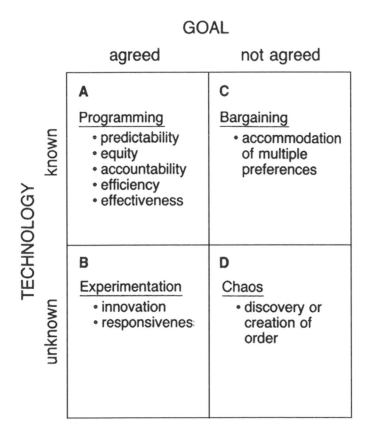

K. CHRISTENSEN, FROM "COPING WITH UNCERTAINTY IN PLANNING", 1985. GOVERNMENT EXPECTATIONS ASSOCIATED WITH PROTOTYPIC CONDITIONS AND RESPONSES TO PLANNING PROBLEMS.

Christensen's matrix in fact represents the conditions of the problem: it is composed of four sectors determined by two axes whose variables pertain to the means and ends in keeping with the degree of certainty. Along the horizontal axis the objectives range from conflicted to consensual, and different value systems can be placed in opposition. Along the vertical access the means or technologies vary from known to unknown, available or unavailable. Moving through the matrix, then, one passes from operative conditions of certainty of means and objectives – we all agree and know the available techniques – in which public action "responds to requirements of reliability, efficiency, and efficacy", to conditions of ever greater uncertainty – when the objectives are multiple and conflicting, and the possible means of achieving them are not known or available. In this case, we are looking at a chaotic condition in which the job of the planner consists of making the problem approachable, helping identify and reconstruct it. Technologies clearly represent a problem of available knowledge. What's more, since they are rooted in and produced by social practices (CHRISTENSEN 1985: 63), they are not neutral. Technologies are not exempt from a system of values, and they influence the formation of the objectives themselves.

Between the two extremes of planning and chaos lie conflicted situations in which the means do not constitute a problem (and thus the requirements of action of the planner are, above all, concerned with outlining a framework of possible actions). In these situations the objectives are shared but the means, the techniques, are not available, or not known. In this case the implementation of various actions requires experimentation and innovation. Just consider the efforts made to "save Venice": while consensus regarding the general goals seems obvious, the difficulty choosing techniques, the uncertainties regarding their efficacy, and the lack of experimentation have slowed their implementation for decades, enough to threaten the consensus regarding the general objectives.

Christensen's diagram returns to the definitions of two families of problems already revealed by other authors. When the problem is "well-defined", the ends to pursue and the means for doing so are clear. The problem is "ill-defined", on the other hand, when the ends and means are not known. But the nature of the design problem and the conditions in which it is constructed are often decidedly complex and belong to a third family (according to some, a sub-family belonging to the ill-defined problems): perverse, "wicked" problems[8]. In this case, the ends to pursue do not become clear until the moment they have been achieved, if they are ever achieved. The exercise permits multiple solutions and therefore it is never resolved; potentially it has no conclusion. Often, Christensen points out, one proceeds with "premature programming", convinced of the existence of a "premature consensus", when the theme should be recontextualized in terms of both the means and the goals the community sets for itself[9]. Most importantly, different degrees of uncertainty should correspond to different styles of planning, different families of projects.

The construction of scenarios comes to terms with this thesis, and in particular with certain parts of Christensen's matrix: those concerning the fields that contain the situations of greatest uncertainty, confronting this condition rather than avoiding it, seeing it as a constituent part of the project instead of a cumbersome burden and obstacle. The scenario, then, is proposed here as "project form", one of the different forms that goes into the design of the city and the territory (VIGANÒ 2000b). It interprets and utilizes the conditions of uncertainty, risk, and fear as materials for its own construction.

Risk.   In mathematical terms, risk is equal to the probability that event *x* will take place; "it is the impossibility of making probability predictions that distinguishes uncertainty from risk" (DRYZEK 1987: 31). According to several scholars, including Scott Lash and Ulrich Beck, the rupture inside modernity as it emerges from the industrial society is linked to a new configuration, the industrial risk society. We are at the start of a new modernity, and in this advanced modernity "the social production of wealth is systematically accompanied by the social production of risks" (BECK 1992: 19). From the problem of the distribution of wealth, at the center of the concerns of Marx and Weber, we pass to the question of the governing of risk – the latter being linked to a process of modernization and completely internal to it.

Often risks are not perceptible or capable of being experienced; to be recognized, they require the "sensory organs of science". But once the risk emerges, it eludes that monopoly, becoming a social and cultural problem that is often disconnected from the scientific validity of the surrounding convictions and from a rigorous mathematical definition. On the other hand, even the scientific statements on risk actually "contain statements of the type *that is how we want to live*" (BECK 1992: 58), overstepping the bounds of the scientific field and superimposing themselves on the aspirations and desires of the society. Every time we construct a scenario, we implicitly or explicitly compare it with a series of *"that is how we want to live"* statements.

Again on risk and rupture: in industrial modernity the increase of productivity is generally seen as the consequence of the process of division of labor, while "risks display an encroaching relation to this trend" (BECK 1992: 70). Risks exist across disciplines, classes, genders. This transverse nature contributes to create a split in the scientific apparatus that come from the 19th century and is based on the idea of specialization and separation. There is no outside to which to deliver the excess or the consequences of estimated risks. The end of the nature/society antithesis is also the end of their separation, of nature as other than society. The sociology of risk is, in Beck's writings, a science of possibility and judgments regarding probabilities; risk is a virtual reality, something non-existent that projects itself on the present and influences it. Like calculations of probability or accident simulations, "risks are related directly and indirectly to cultural definitions and standards of a tolerable or intolerable life" (BECK 1992). Statements of risk lie between virtuality and the non-existent future; they strain, or subject to effort, the structure of the modern edifice, forcing it to absorb certain epistemological and pragmatic changes. The scenario is mainly devoted to this: it constructs non-existent situations that are projected onto the present and can influence its future.

Fear.   The scenarios of Herman Kahn[10] are particularly representative in this sense. Reviewed a few decades later, they are striking for their depiction of the expectations and, above all, the fears of the western world (the United States, but also elsewhere) at the start of the 1960s. After having published a book on the consequences of a thermonuclear war that was appreciated by some but strongly criticized by others, Kahn writes the introduction to *Thinking about the Unthinkable* (1962), recalling the trade in "white slaves", the English girls forced into prostitution, which still in the Victorian era provided for the brothels of all Europe. The hypocrisy and shameful silence that surrounded and permitted this violence constitutes the main subject of the first chapter of the book, with its suggestive title: "In defense of thinking", supporting the necessity of thought and speech regarding shocking topics like the possibility of a nuclear war. "Unthinkable" does not mean *impossible*, and "failure to think may even make it more probable", writes Kahn, regarding nuclear catastrophe[11]. The scenario is one of the "strange aids to thought"[12], like abstract models that simplify but also permit interesting approximations, games, and simulations. A scenario "results from an attempt to describe in more or less detail some hypothetical sequence of events. Scenarios can emphasize different aspects of 'future history'", (or history of the future, in quotations marks in the text). "The scenario is an aid to the imagination".[13] In the long narratives of the "peace and war games" (that remind us of the screenplays of many films during the Cold War years) of Kahn, a member of the Rand Corporation, founder and director of the Hudson Institute, the scenario as script and the scenario as system analysis are joined back together.
The question raised by Ulrich Beck has to do with the consequences, which perhaps we do not fully know how to assess, of the passage from the fellowship of hardship to that of fear, the cohesive force of fear.

Forecasts, conjectures, scenarios

A scenario is not a forecast, or, as I will discuss below, it is a particular form of forecast. The return of interest in the future contains a sense of the passage from the modern idea of *forecast* as projection, connected with the certainty of objectives and the means to achieve them, to that of possibilities and prospects, in conditions of uncertainty, both with respect to objectives and to the consensus around them, and to the knowledge of the means for their achievement. From the idea of the certain or the probable to that of the possible: the future is part of the idea of action.

Forecasts.   If we assign a non-deterministic meaning to the term *forecast*, seeing it as "a substantiated statement regarding a future situation" (MARBACH, MAZZI-

OTTA, RIZZI 1991: 3), since foreseeing the future is a complicated operation, then the scenario can be traced back to the *what…if…?* scheme, and represents a selection of possibilities that accounts for history, as well as for its being subject to modification. Seen in these terms, the forecast is clearly not a prediction, but a hypothetical construction that responds to the need for guidance of action in situations of a deficit of explanations and with an "incomplete experimental legacy" (MARBACH, MAZZIOTTA, RIZZI 1991: 6) – situations in which we are embroiled on a daily basis.

A forecast is one of the ways "to link the present to the future"[14] through the analysis of quantitative data that are often already structured inside projections well prior to the start of the project. In this case the designer loses the chance to manipulate the datum as a substantial part of the construction of scenarios and "visions"[15]. Actually, the selection of the variable whose change is to be analyzed, or of the conditions inside which certain changes can produce themselves, is already a fundamental part of the hypothetical investigation.

Forecasting methods can be classified in many ways: from the viewpoint of the user, by type of asset, period of reference, life cycle, and so on. The approaches can be based on extrapolation, involving observation of the past and present through which to look to the future; or simulation, to evaluate the effects of alternative initiatives; while normative approaches investigate strategies to bring about a given set of circumstances. In this case the scenario moves from the desired future situation towards the present, defining a field of acceptable variation.

One of the best-known scenario exercises that starts with the future and reconstructs all the necessary steps to arrive there is the novel *Looking Backward, 2000–1887* by Edward Bellamy, published in 1888. The protagonist wakes up after more than 130 years in the Boston of the year 2000, and the literary device enables the author to reconstruct the passages that have permitted the individualistic civilization of the late 1800s to be transformed into a collectivist society, a nation that has abolished money and is organized as an army of workers. More than the images of a city of glass for this fantastic and magnificent vision of the Boston of the year 2000, it is the detailed description of the everyday life of the future and its space (obviously as compared to that of the end of the 19th century) that sheds light on all the concerns that will form the basis of the project of modern urban planning. The age of individualism is contrasted with that of the "concert", "of mutual agreement", and the umbrella, an individual solution for shelter from the rain, is replaced by the large covered spaces of the future city of the 20th century, a collective solution to the same problem. The dwellings are small but functional and comfortable, and they extend into the buildings of the people that are great common dining halls and places of recreation and pleasure in which each family has its own space, and in which collective wealth is represented. The limitations of a society exclusively focused on the individual are mercilessly revealed by economic crisis, the trigger factor of the radical transformation narrated by the doctor who reawakens the protagonist. It is a crisis that seems to announce that of 1929 and even the present crises, due not only to overproduction but also to the perverse functioning of the financial machine and systems of credit, which become "a sign of a sign", as Bellamy writes, pressuring the society to develop new systems of solidarity and values.

Conjectures.   While the projection is an extrapolation of phenomenal and political tendencies that is subsequently corrected by hypotheses of trend modification, the forecast is not even the most probable prefiguration, since it is not possible to discern which is the most credible configuration, but only to describe the present. Instead, it is an opinion or, as Jouvenel writes, a reasoned conjecture about the future.

Bertrand de Jouvenel, author of *L'art de la conjecture* and founder of the *Futuribles*[16] project, uses the term *conjecture*, taking it from Jacques Bernoulli[17] but divesting it of the character of probability distribution attributed to it by the latter. Jouvenel reconstructs the debate regarding the study of the future that took place in the 18th century between Maupertuis, a member of the Academy of Sciences of Paris, and Voltaire. Establishing a symmetry between memory and forecast[18], Maupertuis prompts reflection on the imperfection of both the reconstruction of the past and the forecasting of the future: knowledge of the past is just as uncertain as knowledge of the future, but both can be perfected thanks to the advances that happen in the different fields of science. Vast parts of the history of the past and also of the future have yet to come to light, and this makes it possible to avoid errors, such as that of Montesquieu later corrected by Hume on the trend of depopulation of the earth, based on erroneous knowledge of the numbers of the past.

Voltaire, on the other hand, thinks the symmetry between memory and forecast is utterly unreasonable, and above all sees the latter term as unsuitable to human knowledge. In effect, consulting the dictionaries contemporary to this *querelle*, Jouvenel makes a comparison between two terms: *prévision*, knowledge of what is to come, and therefore a divine faculty, and *prévoyance*, "action of the spirit that considers what can come" (JOUVENEL 1964: 28) — not what will happen, but the array of possible futures. He also emphasizes the caution required in the modern use of *prévision*, which must not give rise to ambiguities caused by the lingering of the "scent of its classic, strong meaning" of full knowledge of the future (JOUVENEL 1964: 29). On the other hand, the definition of *prévoyance*, "foresight", but also "far-sightedness", involves "the foreseeing of future cases and the taking of the appropriate measures"[19], as if the human application of forecasting were simply a matter of concern for the material bases of survival in the future, rather than one of abstract speculation. I think this is a pertinent point that clarifies the reasons to do research and a project through the construction of scenarios. This is also the reason why Jouvenel rejects the terms of "science of the future" or that of "futurology" coined in the mid-1940s by Ossip K. Flechtheim. In an essay published in *The Futurists* (TOFFLER 1972),

Flechtheim actually seems to be close to the philosophical questions raised by the French school and to share Jouvenel's doubts: though he introduced the term *Futurology*, he does not make claims regarding its scientific tenability, considering it "either as a science or as a 'pre-scientific' branch of knowledge", depending on the definition of science used. For Flechtheim, futurology is "a projection of history into a new time dimension"[20], but its methods differ from those of historical investigation, and are closer to those of the social sciences, from which it is separated by the investigation of mathematical probability and "credibility". The latter is a wider-ranging concept, but it can be rendered more precise via forms of reasoning that are both inductive and deductive – here the reference is to Bertrand Russell. Published in the same year as *The Limits to Growth*, Flechtheim's essay seems to share its millennium angst: "Finally, in our days a clear-cut and unequivocal picture of the future could turn into a sublime personal challenge to those who are ready to withstand the inevitable with courage and conviction"[21]. Futurology is concerned with the fate of civilization.

Conjecture is "the intellectual construction of a plausible future", a true "work of art" (JOUVENEL 1964: 31), a logical structure that is transparent in its assumptions and statements. The term *futurable* taken up by Luis de Molina, a Spanish Jesuit in the 16th century, reduces the range of future possibilities, combining the terms *future* and *possible*: states are futurable if "their mode of production is imaginable and plausible for us in the present" (JOUVENEL 1964: 32)[22]. The reduction of the field of possibilities to only those we are already capable of producing does not, in any case, resolve the impossibility of an exhaustive representation of a future which continuously varies in time and space.

Scenarios.   Scenarios are systemic methods[23], investigations of the possible lines of evolution of given phenomena, and they can also follow diverging trajectories, developing contrasting hypotheses and images. Often the effort of imagination necessary for the construction of scenarios seen in this way produces simple alternatives that act on a rhetorical level and can easily exclude each other. Though it is clear to all that the simulation of scenarios will not automatically and in linear ways be transmuted into a blueprint without further thresholds of reformulation, the main interest in their use seems to be that of allowing us to free ourselves of doubts and uncertainties regarding the future and stem the anxiety that grips us. We rarely wonder what would happen if different scenarios were to coexist, accumulating frictions but also new and possible relationships, mutual disturbances, or even point to unexpected inter-relations. In reflections developed during the course of the construction of the territorial plan of Salento, the overlaying of scenarios revealed new themes of the design of the territory that spring from imagining possible and unexpected coexistences.

Constructing reflections about the future, a "procedure capable of both creating images of the future and, above all, of systematically making the possible consequences of certain situations emerge", hypothetical sequences of events, "prefiguration of hypothetical situations, internally coherent within explicit hypotheses" (MARBACH, MAZZIOTTA, RIZZI 1991: 33): this is the important role the scenario can play in the reconfiguration, today, of the design of the city and the territory.

[1] I borrow the apt title of the lecture by Dominique Rouillard at the *III International PhD seminar, Urbanism and Urbanization*, Università IUAV, Venice, 2006.

[2] Commission on the Year 2000, Toward the Year 2000: work in progress, 1967.

[3] The term comes from the sociologist Daniel Bell, 1967: "Notes on the Post-Industrial Society", in *Public Interest*, Winter and Spring.

[4] Cited by Armatte, *Les économistes face au long terme: l'ascension de la notion de scénario*, in DAHAN DALMEDICO 2007: 65.

[5] The first study for the presidential commission formed by President Hoover and financed by the Rockefeller Foundation dates back to 1929, done among others by W. Mitchell and W. Ogburn, concluded in 1933, with the title *Recent Social Trends in the United States*.

[6] A figure: "souhaitable (politiquement et socialement), praticable (c'est-à-dire réalisable) et probable (comme prolongement du présent)" (MASSÉ 1965).

[7] The term is introduced by Gaston Berger in 1957. See the second chapter in this same part.

[8] WEST CHURCHMAN 1967; RITTEL, WEBBER 1973; ROWE 1987; COYNE 2005.

[9] The risk of interpretation of the urban planning project as a banal machine is always very strong. See SECCHI 1988.

[10] Among them: KAHN 1962; KAHN, WIENER 1967; KAHN, BRUCE-BRIGGS 1972.

[11] KAHN 1962: 39.

[12] "Strange aids to thought" is the title of the fifth chapter of *Thinking about the Unthinkable*, KAHN 1962.

[13] KAHN 1962: 143.

[14] HOPKINS, ZAPATA 2007: 4.

[15] In particular Isserman A.M., *Forecasting to Learn How the World Can Work*, in HOPKINS, ZAPATA 2007.

[16] JOUVENEL 1964. The "Projet Futuribles" proposed by Jouvenel in 1960 is financed by the Ford Foundation. *Futuribles* is the title of the magazine founded and directed by the son of Jouvenel in 1974, containing essays on future studies.

[17] The reference is to the work *Ars Conjectandi*, published in 1713 after the death of Bernoulli. *L'art de la conjecture*, for Bernoulli, is a stochastic art that assesses probability.

[18] "L'une est un retour sur le passé, l'autre une anticipation de l'avenir", Maupertuis, P. L., *Lettres* (orig. ed. 1752), *Lettre II*, tome II of the Complete Works, ed. 1768, pp. 222; cited by JOUVENEL 1964.

[19] *Dizionario di Italiano*, 2003, Garzanti Editore.

[20] FLECHTHEIM 1972.

[21] FLECHTHEIM 1972.

[22] In 1588, Molina publishes a work on free will. A futurable is "un descendant du présent qui comporte une généalogie" (JOUVENEL 1964: 32).

[23] The term *scenario* in the sense utilized here is introduced by Kahn in *Thinking about the Unthinkable*.

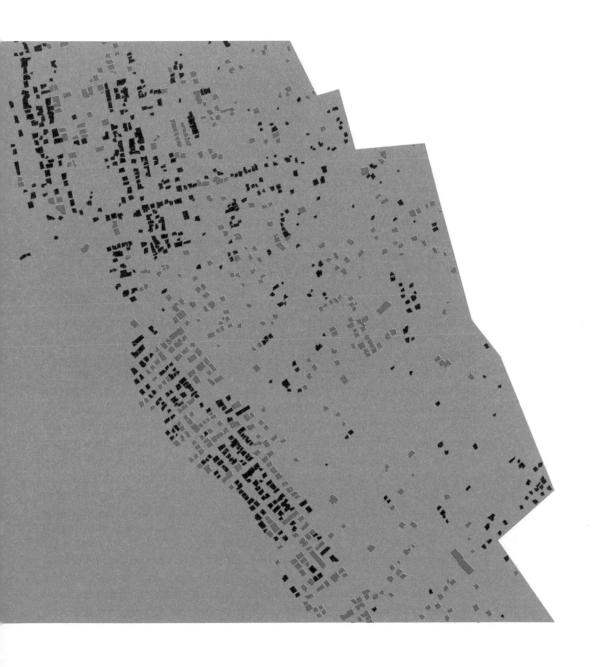

One of the main phenomena characterizing the Province of Lecce in past years is the dispersion of residential and productive settlements in the agricultural territory, a phenomenon that is difficult to assess both in its quantitative aspects and in its relationship to urban planning and environmental standards.

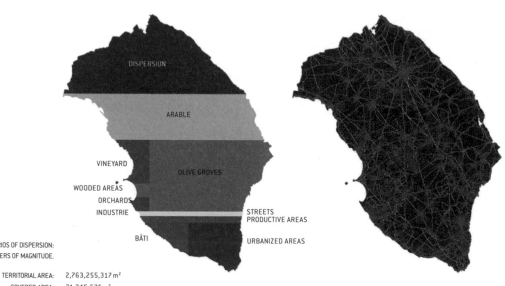

DISPERSION

ARABLE

VINEYARD

OLIVE GROVES

WOODED AREAS
ORCHARDS
INDUSTRIE

STREETS
PRODUCTIVE AREAS

BÂTI

URBANIZED AREAS

SCENARIOS OF DISPERSION:
ORDERS OF MAGNITUDE.

TERRITORIAL AREA:	2,763,255,317 m²
COVERED AREA:	71,245,575 m²
GENERIC CONSTRUCTION:	62,174,865 m²
INDUSTRIAL CONSTRUCTION:	3,049,258 m²
AGRICULTURAL CONSTRUCTION:	84,698 m²
GREENHOUSES, CANOPIES, SHELTERS:	4,644,707 m²
POPULATION:	817,398 (ISTAT 1998)
DENSITY inhab/km²:	298 (ISTAT 1998)

IN RED, CONSTRUCTION
MORPHOLOGY (TO THE RIGHT).

P. VIGANÒ (ÉD.), *TERRITORI DI UNA NUOVA MODERNITÀ / TERRITORIES OF A NEW MODERNITY*, NAPOLI: ELECTA.
THE TERRITORIAL PLAN OF THE PROVINCE OF LECCE WAS PREPARED BY A LARGE GROUP, COMPOSED AS FOLLOWS: P. VIGANÒ (DESIGNER), B. SECCHI (SCIENTIFIC CONSULTANT), S. MININANNI (COORDINATOR *STUDIOLECCEPTCP*); S. ALONZI, L. CAPURSO, A. F. GAGLIARDI, A. D'ANGELO, L. FABIAN, R. IMPERATO, F. PISANÒ, M. D'AMBROS, R. MIGLIETTA: *STUDIOLECCEPTCP*; C. BIANCHETTI WITH P. DE STEFANO, G. PASQUI, L. VETTORETTO (LOCAL DEVELOPMENT POLICIES); M. MININNI WITH S. CARBONARA, P. MAIROTA, N. MARTINELLI, G. CARLONE, G. MARZANO, L. SCARPINA, P. MEDAGLI, L. ROSITANI, M. LAMACCHIA, D. SALLUSTRO (ENVIRONMENTAL AND LANDSCAPE ASPECTS); A. TOMEI (GEOLOGICAL AND HYDROGEOLOGICAL ASPECTS); A. DE GIORGI (ALTERNATIVE ENERGY POLICIES).

If the trends of the years from 1961 to 1991 were to continue unchanged over the next twenty years in Salento, about 150,000 new residences would be built. If the increase in living space of each lodging were to continue, as a result of higher levels of income and quality of life, and associated with the preference for different building types than in the past, with more single-family homes scattered in the countryside, about 15 million square meters of living space would be added to the present quantity. Part of these dwellings and this area would be scattered in the countryside. Obviously these are approximate estimates, but they convey an idea of the order of magnitude of the phenomenon.

New neighborhoods or houses with yards?   Around the centers: the new dwellings could, for example, be built with the settlement principles, building types, and densities of one of the many public development initiatives (Peep) of the Province of Lecce. In this case, these would be neighborhoods composed, on average, of three-story buildings in orderly arrangements, with about three quarters of the land area set aside for equipment and greenery. They would make use of about 21 million square meters of area, and could be located around or near urban centers.

The settlement profile of Salento itself would undergo limited change.

If, instead – in an extreme and opposing scenario – all the new dwellings were single-family homes scattered in the countryside with the ratios of lot coverage typical of many zones of the urbanized countryside of Salento, these houses would occupy over 700 million square meters of area. This is an enormous area that would radically alter the territory, giving rise to a different city.

The new construction could be arranged around the centers (as is already happening), or along the main axes of communication (less probable), or along the coasts (as has already taken place in many parts), and this would raise problem of no small impact, mostly connected with the water cycle. These are extreme scenarios; the new construction will not follow a single settle-

1

2

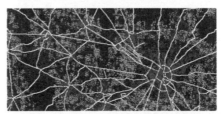

3

4

HYPOTHESES 1, 2, 3, ON THE DENSITY OF THE URBANIZED COUNTRYSIDE.
HYPOTHESIS 4, ON THE DENSITY OF THE PEEPS AND AROUND THE CENTERS.

ment model. Still, it is interesting to think about possible directions and modes of development, even if they would not be probable on this scale.

Recovering unauthorized urbanization along the coast?   What would happen if the strategy for recovery of unauthorized areas involved their progressive densi-

TODAY

INHABITED AREA, AS OF 2021
61,210,745 m²

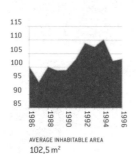

AVERAGE INHABITABLE AREA
102,5 m²

2021/A

SCENARIOS OF DISPERSION: TRENDS IN PROGRESS. THE TENDENCY OF INCREASE OF INHABITABLE AREA FROM 2001 TO 2021 IS 15,282,210 m² THE INCREASE OF NUMBER OF DWELLINGS FROM 2001 TO 2021 IS 149,124.

[TO THE RIGHT] PRESENT STATE. SCENARIO OF DENSIFICATION AROUND THE NUCLEI AT THE DENSITY OF THE PEEPS. SCENARIO OF DENSIFICATION AROUND THE NUCLEI AT THE DENSITY OF THE URBANIZED COUNTRYSIDE.

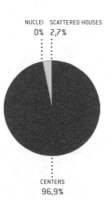

NUCLEI   SCATTERED HOUSES
0%   2,7%

CENTERS
96,9%

2021/B

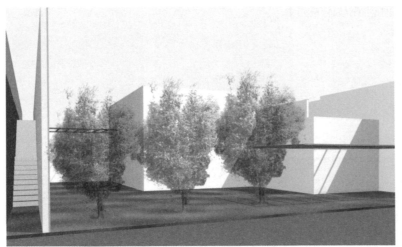
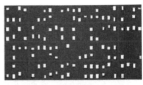
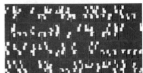

fication? Would this prevent construction in other areas? What landscape, what costs, and what advantages would this bring to the community? Would it imply centralized decisions?

An initial scenario imagines that every owner can increase the density of his own lot, continuing the spontaneous process of construction. The first phase would be followed by two other phases: one to construct a residence for the "children" and one for the "grandchildren", adding the missing infrastructures. The resulting landscape would be both compact and porous.

The second scenario is based on the hypothesis of reconstructing the environmental resources unauthorized development has modified or destroyed, demolishing the first band of houses on the sand dune. This scenario implies an overall design: the design of open space guides the recovery intervention, defining connections, interruptions, and infiltrations.

A third scenario also exists, which imagines demolishing the unauthorized areas built along certain segments of the coast. This implies different policies and operators than those of the present, who know how to build strong consensus around certain themes and solutions of common interest.

Reasoning through scenarios makes it precisely possible to evaluate the conditions in which a society is willing to make certain choices, for example to recover unauthorized zones and the most important constituent elements of an environment the society appreciates.

Countering desertification.   A territory covered only up to 5.5% by semi-natural and sub-natural vegetation cannot combat the phenomena connected with global change, climate changes, desertification, and generally irreversible changes; and it does not contribute, in particular, to the reduction of CO_2.

VIEWS OF OPERATIONS
OF DENSIFICATION AND
INFRASTRUCTURING OF ZONES OF
UNAUTHORIZED DEVELOPMENT
WITH NEW PUBLIC SPACES.

OVERLAY: 311,571 m²
LANDED PROPERTY AREA: 216,000 m²
EMPTY LOTS : 77,400 m²
GARAGES
FIRST GENERATION: 100,054 m²
SECOND GENERATION: 204,002 m²
THIRD GENERATION: 297,337 m²

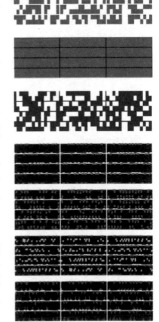

 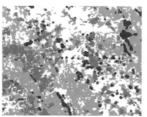 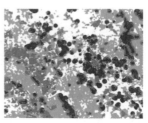

COUNTERING DESERTIFICATION: SCENARIO IN PHASES THAT BEGINS BY IDENTIFYING BUFFERS OF POTENTIAL EXPANSION OF EXISTING VEGETATION AND EVALUATES THEIR POSSIBLE SPREAD IN RELATION TO AGRICULTURE AND OCCUPATION OF THE LAND.

Starting with the present environmental mosaic, assigning a significant role for the expansion of naturalness to all its components (houses, streets, hedges, gardens, waters, vegetation, and phytoclimate) and considering the agricultural component as the matrix (in the etymological sense of the term) of the landscape, it is possible to construct a scenario of expansion of naturalness.

The construction of the scenario makes use of a recursive procedure, in phases:

phase 1: construction (starting from combinatory spatial rules that indicate number, form, contiguity/isolation, homogeneity/heterogeneity of natural areas) of buffers, i.e. areas of potential expansion of existing vegetation;

phase 2: evaluation of the capacity of the environmental matrix to permit or to intervene as a barrier to expansion and spread of naturalness in keeping with: a) agriculture, its structuring and its productivity; b) settlement models of dispersion/concentration of housing and their occupation of the land area; c) strategies of safeguarding of land and water;

phase 3: repeats phase 1, using the results obtained after phase 2 as its starting point;

phase 4: repeats phase 2, starting with the results obtained after phase 3.

Conflicting scenarios

Scenarios have to be properly understood. They do not constitute a forecast, but rather an attempt to grasp what would happen were certain events or phenomena to take place. Many of the scenarios we have constructed are rooted in trends currently in progress:

1. The scenario of expansion of the vineyard is based on observation of the growth not yet fully recorded by maps of areas cultivated as vineyards.

2. The scenario of habitation has been developed in keeping with two fundamental modes: living in neighborhoods endowed with a certain urbanness, near city centers, or living in the country in houses with yards placed along the coast or in the inland zones.

3. The scenario that refers to productive and commercial activities has also been developed according to two fundamental modes of concentration, inside specially earmarked areas or dispersed along the main roads that cross Salento.

4. The scenario of expansion of naturalness has been constructed in a different way. It clearly demonstrates how a porous territory could function if the expansion of naturalness is not obstructed.

5. These scenarios are joined by the energy scenarios, which represent an initial attempt to grant a dimension to the energy problem of Salento and to the potentialities of this territory.

Some of these scenarios have a clear character of contrast or repulsion; they point to undesirable consequences. Others leave room for greater possibilities of consensus; they demonstrate that things could progress in a certain way as long as certain requirements are fulfilled. Others still appear to be desirable and

ELEMENTS CONFLICTING WITH THE EXPANSION OF NATURALNESS. SCENARIO OF ADJUSTED EXPANSION OF NATURALNESS (ON TOP).

ELEMENTS CONFLICTING WITH THE SCENARIO OF DENSIFICATION OF THE PLAINS AND ZONES OF PROXIMITY (HIGH RISK OF FLOODING, ROOMS OF THE PARK).
ELEMENTS CONFLICTING WITH THE SCENARIO OF EXPANSION OF DISPERSION (STRIPS OF SAFEGUARDING AND REPLENISHMENT).
ADJUSTED EXPANSION OF DISPERSION, PLAINS, AND ZONES OF PROXIMITY TO INHABITED CENTERS (CENTER).

ELEMENTS CONFLICTING WITH THE SCENARIO OF EXPANSION OF INDUSTRY (HIGH RISK OF FLOODING, URBANIZED AREAS, ROOMS OF THE PARK).
ADJUSTED SCENARIO OF EXPANSION OF INDUSTRY.
ELEMENTS CONFLICTING WITH THE SCENARIO OF VINEYARD EXPANSION (BOTTOM).

ADJUSTED SCENARIO OF VINEYARD EXPANSION.

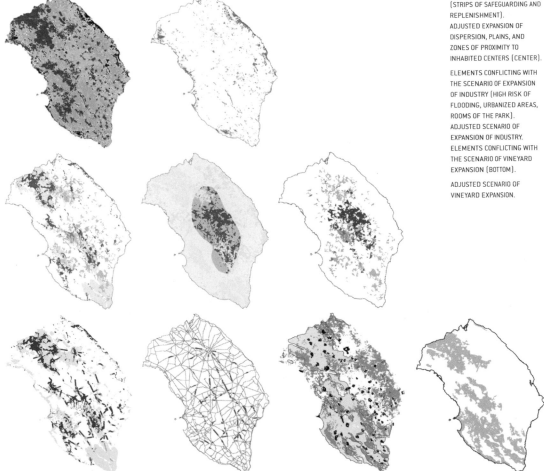

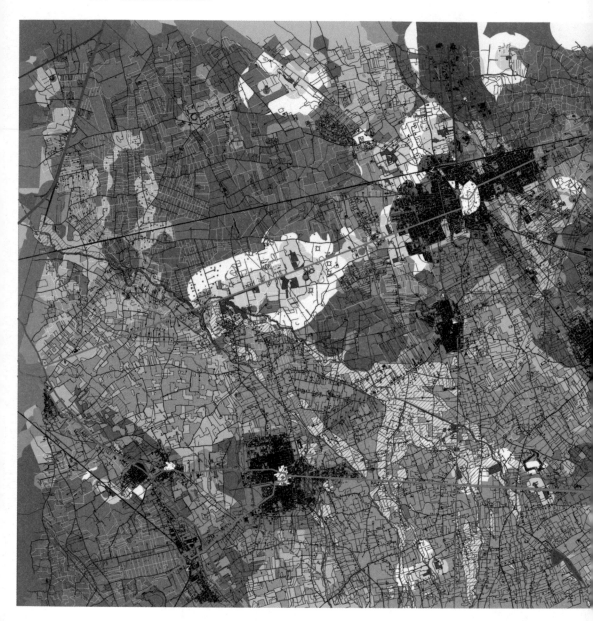

TERRITORIAL COORDINATION PLAN
OF THE PROVINCE OF LECCE,
ORIGINAL SCALE 1: 25.000
(EXCERPT).

appealing. Other scenarios could be superimposed on these, for example: scenarios of concentration of commercial activities along certain axes, scenarios of decentralization of collective facilities to form a new habitat of accessibility, scenarios of recovery of historical centers and of the *masserie* as the biggest hotel in the world.

The map is enriched by layers and overlays, demonstrating both the need to make choices and the possibility of integrations.

Compatibility and incompatibility: new coexistence.   The construction of each scenario has made use of different information, data, and methods. The set of the different scenarios is therefore an overlay of heterogeneous objects.

The overlay of different scenarios suggests two possibilities: their reciprocal compatibility or their incompatibility.

Incompatibility prompts selection, for example to choose between expansion of the vineyard and the spread of housing. Choice means political construction, and this has to be expressed through rigorous projects.

Consistency and compatibility, on the other hand, indicate possibilities of integration of different scenarios, opening in general to original and innovative design themes. They require construction of policies and projects that act on different levels: for the integration of dispersed housing and naturalness, for example, or for the spread of naturalness and the expansion of areas set aside for special forms of agriculture. This is the reason behind the need for careful observation of the overlapping areas of the different scenarios, where the possibility of a project of new coexistence emerges most clearly.

NARRATIVE ITINERARIES.

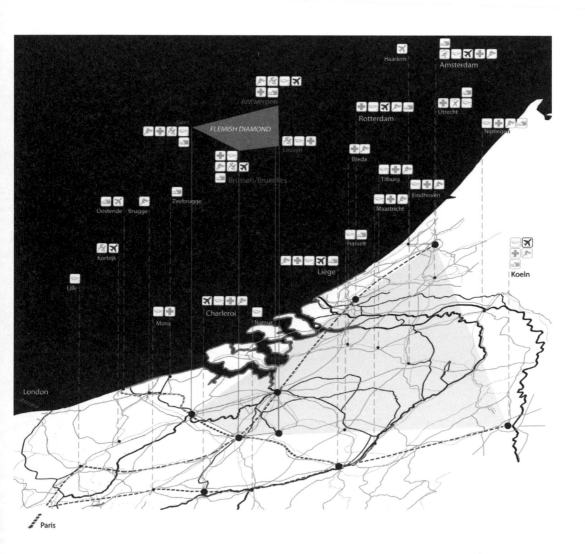

Haarlem

Amsterdam

Antwerpen

Rotterdam

Gent

FLEMISH DIAMOND

Utrecht

Leuven

Nijmegen

Breda

Brussels/Bruxelles

Tilburg

Eindhoven

Maastricht

Oostende Brugge

Zeebrugge

Hasselt

Kortrijk

Liège

Koeln

Lille

Charleroi

Mons   Namur

London

Paris

Phenomenology of the future

Gaston Berger, translator and interpreter of the phenomenology of Husserl, introduces the term *prospective* in 1957: *prospective* (to look forward) as opposed to retrospective (to look back). "*Regard*" is what we cast back towards the past; "*projet*" regards the future, in which possibilities open up[1].

Time and *prospective* are terms approached by the philosopher, who passed away in the same week in which he was to begin his course entitled "*Precurseurs et prospective*" at the École Pratique des Hautes Études. The first lecture, of which a manuscript remains outlining the main points, is entitled: *Phénoménologie du temps*[2]. Berger underscores the appearance of the theme of time, the time of evolution, the historical times at the center of the concerns of the 19th and then the 20th century. The focus on time foregrounds the present, a "thick" present in which "we have the experience of what is to come"[3]. While today it is the perception of global ecological risk that has pushed the future back into the center of our thoughts, for Berger, in the 1950s, it seemed instead that the acceleration of history was leading the interest of society towards the theme of the future and its construction. What the scenario portrays is not "the unfamiliar and rapidly changing world of the present and of the future"[4]; research on the future is not only the construction of alternatives, or an accurate method of studying possibilities other types of analysis might not glimpse, bringing them to attention. The study of the future is, above all, a philosophical problem that, through Berger and other authors, permits the reconstruction of reflections of great interest on the scenario, a relevant point to fully grasp the meaning of the design operation I am focusing on in this part of the book.

An open conception of time, an exploded time that presents itself in its making, and that gets away from habits and conventions, according to Berger, is the great philosophical contribution of Bergson[5]. The rapid transformations of the world emerging from World War II make it possible to concretely grasp the Bergsonian *durée*, in which "one incessantly implements a radical remaking of everything"[6]. While for some the renunciation of a stable, repetitive idea of time produces anguish, the opening towards new representations of the world and the idea of possibility define a backdrop open to the idea of action and the future. The future appears as a vast territory to be colonized.

SCENARIO: *LIVING IN THE MEGASTAD, ANTWERP, TERRITORY OF A NEW MODERNITY,* BY SECCHI B., VIGANÒ P. 2009.

The modern idea of "open time" makes a radical break with previous conceptions of time, with knowledge of the future being but one portion of a wider awareness of time that makes reference to a given place and a given civilization, to a precise state of language and culture[7]. Pierre Bertaux, who writes together with Berger in the *Encyclopédie française*, asks himself about the capacity to perceive this "temporal field" without a verbal form that refers to the future, as was the case of ancient Hebrew, to the point of imagining that the invectives of the prophet Isaiah directed at a people that did not follow him and listen to his predictions were evidence of the inability to conceive of a temporary dimension that belongs neither to the past or to the present. The idea of the future has only interested philosophers in recent centuries: Berger points to the absence, in the *Encyclopédie* of Diderot, and in the *Dictionnaire philosophique* of Voltaire, of the term *future*, which instead appears in the *Grande Encyclopédie* of the 19th century, albeit with a very limited meaning[8]. Even in fields and periods closer to our own, the lack of interest in the future on the part of the academic universe of planners, caused, according to Myers and Kitsuse (2000), by the centrality of spatial analysis and the overwhelming presence of the social sciences (that accustom us to analyzing only what can be traced back to a datum, relative to a past or present phenomenon), produces – in the American context examined by the two authors – a thorny contrast with the professional world, increasingly urged by politicians and inhabitants to prepare "blue sky wish lists" that do not rigorously approach the discussion about possible futures.

As a phenomenologist, Berger does not seek the essence of the future, or what it is in itself, "but the meaning we attribute to it, both explicitly in our discourse and implicitly in our acts"[9]. Thus it becomes important to distinguish between *future* – that which has to happen – a destiny that awaits us, already formed in advance, and *becoming* – what is to come – which is also the result of what we are capable of constructing. The positivist worldview – with its faith in the foreseeability of things, in the possibility of extending the past into the future[10] – leaves room for an open idea of time in which the hypothesis of the "inversion of time"[11] takes on meaning, as formulated by Berger interpreting the path of humankind as a deep mutation, that resembles a new birth instead of a progressive ageing. An admirer of the scientist, paleontologist, geologist, and priest Teilhard de Chardin, Berger affirms the conviction of the Jesuit philosopher that humanity, far from being aged at this point, is instead a young species just taking its first steps.

The reflection on time is central to moving beyond the disjunction between present, past, and future, and to shifting from a "phenomenology of acceleration " to a "metaphysics of hope"[12]. The greatest effort has to deal with the major shortcomings, the lack of imagination, the weakness and meagerness of our capacities. This is why we often rely on elementary devices, developing simplified approaches to the future dimension that, in European culture, passes through space. We construct symmetries, we overturn the past into the future, we extend the temporal experience already crossed into the future, or we extrapolate certain trends from the present. The limits and risks of the three approaches – symmetry, extension, extrapolation – are clear.

Awareness of not being able to rely on past tendencies leads Berger to critique the usual modes of constructing decisions: relying on a precedent, for example, represents on a juridical level a *"quasi-contrat"*[13] that commits a social group to not change its judgment and to reinforce the status quo; reliance on an analogy means marked simplification of complexity and the relationships between phenomena; extrapolation, finally, is simply the prolonging into the future of trends whose law of development has already been recognized. According to Berger, these are automatic procedures that presuppose a stable world and have the sole purpose of avoiding, in every instance, the pursuit of active thought about the future. Next to these procedures that are useful only for initial explorations, Berger positions the *prospective*, which addresses the theme of the general conditions in which humankind will find itself in the future, a game whose rules change during the course of the play, along with the number and properties of the single pieces. Unlike the previously outlined procedures, the prospective seeks the deeper intentions that have generated them, starting with the forms of the various phenomena. Although at times it isolates the different factors, it asks itself what would be the consequences of their evolution, then reunites them in a general reflection in which the various consequences are made to interact.

The unknown dimension of the future prompts a conceptual construction similar to Renaissance perspective. It simultaneously constructs, conceptualizes, and theorizes space. Thus is the prospective a construction in time, its conceptualizing, and its theorization. It has to lead to a new worldview, an in-depth vision that "in this case will no longer be in space, but in time"[14]. It constructs itself in the terms of and in keeping with a *"milieu-temps"*.

The rapid change observed in society at the end of World War II, a phenomenon the prospective investigates and that is announced by the convergence of many factors, is on par with a biological change, with the appearance of a new man that will force an exit from the cradle of humanism. Here, again, we find the emphasis on the idea of a new human type that was central to the project and the city of the Modern Movement. Although Bernardo Secchi places the thrust for renewal at the end of the terrible experience of World War I and questions its regenerative capacity after World War II, I can recognize, in the rich accumulation of works on the study of the future during the course of the 1950s, the epigones of that "great generation"[15] that had thought of the new man as the end and reason of an epochal renewal. Many of these – above all on the French scene – come from the Resistance, are high functionaries or *patrons* of big industry, and often holding Christian beliefs. In this context, the voluntary approach of faith in the ability to construct the future that distinguishes the prospective from the Anglo-Saxon and, more specifically, the American tradition, which is more pragmatic and instrumental, can be even more precisely situated.

The task is to approach change by creating new social structures and a new "état d'esprit"[16] in the study of *situations* (in italics in Berger's text) humankind will have to face: an *"anthropologie prospective"*[17] that requires the joint thinking of philosophers "attentive to the goals and concerned with the values" and specialists in the various fields.

Concrete description of future situations: this is the prospective, seen as the phenomenology of the future.

A theory of change

The "prospective" attitude is summed up by Berger in the *Encyclopédie française*[18] in five points, recalling that Valéry wrote *"nous entrons dans l'avenir à reculons",* an image that returns in the tragic figure of Benjamin's angel of history, inspired by Klee's painting, *Angelus Novus*. The angel, wrote Benjamin, seems to retreat from what he observes and attracts, the past: "His eyes are staring, his mouth is open, his wings are spread. […] Where we perceive a chain of events, he sees one single catastrophe that keeps piling ruin upon ruin and hurls it in front of his feet". The angel turns his back on the future towards which he is pushed by a tempest that, as Benjamin writes, "we call progress"[19].

"To see far" and "to see wide" are the first efforts required. In this paragraph, Berger constructs a succinct map of other research in progress in the 1950s within this new field of future studies; this research led to the creation, in large corporations, of *"départements du futur", "bureaux des hypothèses"*. For Berger, the dynamic of change analyzed in the text on Kurt Lewin by Lippitt, Watson, and Westley (1958) regarding social organization, is a "pioneering attempt to formulate concepts on a comprehensive scale"[20]. The agents of change on different scales are individuals, "face-to-face groups", organizations, and community – dynamic systems that act, prepare, and engender change.

The temporal horizon of the prospective should not be reduced to grazing the contingent, but should extend to the long term. Free of any problem of prediction and interested in situations rather than events, it addresses long-term trends and is complementary – not contradictory – to short-term forecasting. In this area, Berger agrees with Kahn about the importance of the "multifold" long-term trend that is utilized as an "organizational concept"[21]. A concrete description of a future situation, the prospective requires "in-depth analysis" that goes beyond automatic procedures and prompts us to "take risks". The short-term decision inevitably reduces the horizon of the possible, while long-term reasoning leaves us free to imagine, to "think about man" at the center of the prospective reflection.

Berger's five points can be compared to an early text of great interest by H.G. Wells: *The Discovery of the Future* (1902). After the fantastic *Anticipations* of 1901 with its detailed description of the new landscape as "neither city nor country", resulting from the dispersion of settlements in the British territory, and after having published his most famous science fiction novels, Wells tries to make his focus on the study of the future more rigorous. Many futurologists cite the text[22] as an impetus to put research on the future at the center of the social sciences; for Wells, this was the start of the social and political engagement that would lead him to become a member of the Fabian Society.

In *The Discovery of the Future,* H.G. Wells asserts the possibility of constructing knowledge of the future. For this to happen, a convergence of scientific, economic, and social studies around the theme of the future is needed, and it must be constantly and "courageously"[23] kept in play. Like Maupertuis, Wells establishes symmetry between past and future: both are possible objects of knowledge, but the approaches can differ. The continuous reference to history characterizes the "legal" type that seeks confirmation, continuity, and legitimacy in the past. A vision of the world as a gigantic "workshop" characterizes, on the other hand, the "legislative" type, better suited to a period of "extraordinary uncertainty and indecision upon endless questions − moral questions, aesthetic questions, religious and political questions"[24], such as the start of the 20th century. The two approaches also differ in their attitude to change: the former focuses on causes, while the latter is concerned with the probable consequences of actions and phenomena. Finally, the basis for effective thinking about the future is the rational study of history, causality, and major trends[25].

According to Wells, the scientific use of the imagination, an imagination trained in the methods of scientific research to have "a habit of inquiry"[26], is the main force behind advances in knowledge of the world. For example, knowledge of the pre-human, geological past of the earth, made possible by an infinite series of questions and inferences, leads him to speak of an "inductive past" to which an "inductive future"[27] should correspond, of inferences projected forward rather than backward. The knowledge of the future that interests Wells is general. It does not eliminate the problem of individual responsibility, but pushes aside the role of the great leaders who do not have the possibility of modifying the long term in which not even the centrality of the human species can be taken for granted.

While H.G. Wells and Gaston Berger express profound philosophical thought on time and change, Jouvenel − who takes them both as references − puts politics at the center of the research on possible futures and the role of representation in stimulating action. Prior to, and differently from, sensation, representation of the future event is the "start of movement" and therefore of action. The point is pertinent to our discussion due to the role of architects and urbanists as builders of images that contribute to the formation of a wider-ranging collective imaginary. When the images produced address the future, not merely the near future that includes the temporal horizon of the project we are to complete, they contribute to making the future an object of knowledge, discussion, and construction. At the same time, they interpret the desire for change that is increasingly evident and aggressive, according to Jouvenel, and the sense of greater uncertainty that inevitably accompanies it. The greater the progress of change, the greater the anxiety about the future.

Horizons of change.   In prospective scenarios, three fundamental conditions of change can intervene: maintenance of the status quo, gradual changes, and "abso- lute discontinuities" (CAZES et AL. 1993). Within the three horizons of change we can recognize "structural certainties" (JOUVENEL 1964), constants, factors of slow change and of substantial continuity; gradual or tendential changes that, cumula- tively modifying over the observed period, seem to ensure a certain stability; and, finally, discontinuities or ruptures, the weak signals we are often unable to recog- nize and that take us back to the start of this chapter, to Peirce, to abduction and the capacity to guess, sooner or later, the evolution of a given phenomenon.

The three modes of change interact in the construction of scenarios and play different roles within it. Recognition of structural certainties allows us to define the context of the scenario, while trends and hypotheses of rupture define a field inside which contrasting scenarios can be developed. Outside the probable, the scenarios force us to think about change in its different forms and as a material fact[28], the result of hypotheses that are logically articulated and can be described in their concrete aspects.

As Kahn aptly clarifies, two approaches can be taken: the descriptive approach, a sort of "value-free perspective"[29] that attempts to imagine the future in the perspective of a necessary adaptation without immediately applying a value judgment; and a second, normative approach that instead investigates the pos- sibilities of constructing a desired future. If it is true that it is not always easy to separate the two approaches, which often at least partially overlap, it is not legiti- mate to transmit what we *would wish* to happen as a hypothetical construction of the future, i.e. as a scenario[30].

Possibilisme.   What I see emerging from the various contributions and the above reflections is an embryonic theory of change, a hybrid framework that is not a sharply hewn or excessively formal theory. Rather, it is a potentially lucid one as it comes across the many difficulties of the contemporary design of cities and territories: the long term as a possible object of research, the inductive future as a field of exploration, the general concern alongside the local, the focus on con- sequences and not only causes, the production of images that connect the tem- poral and spatial fields. This theory scaffolding finds a fundamental and precious link, an important facet pertaining to the past and its imperfect knowledge, in the *Tableau* of Vidal de la Blache, where descriptive practice becomes "practice of hypothesis"[31].

Implicit at times, the *"si... alors"* hypothetical chains interrogate places and his- tory regarding the possibility that things could go differently, always insisting on the complex relationship between natural support and the artificiality of human transformations. Often it is the term *indice*, or clue, that opens the imagination to possibilities that are certainly remote, such as those linked to the processes of erosion of rocks, but nevertheless are inscribed not just in places but also in the successions of transformations that have impacted them over the very long span of geological time. Clues of processes barely hinted at, or already concluded, which, however, could have continued, offer glimpses of other possible evolutions, other landscapes that non-realization does not totally deprive of the sense of reality and of belonging to a *milieu*. It is a useful exercise to see a territory as a deposit not only of the traces of the events in which it has been involved, but also of those that would have been able to transform it and did not, do so[32].

The scenario turns and looks back, and history is made with *ifs*: "if certain trends, for example, had continued; if the store of information available in each case had been larger, and had guided the choices in a different way from what has effectively happened; if certain images stored in the collective imaginary or represented in certain projects had acted as the guiding factors of the decision-making process [...]"[33]. This viewpoint must be used with caution, as was possible during a short and utterly initial reflection on the territory of Modena, open to possible paths of research that involve the past as much as the future. The presumption is not to rewrite history but to investigate the consequences of actions, choices and policies, alongside their necessity or precariousness, freeing our sense of the past from a deductive linearity that often limits and confines our thinking about the future — as well as that about the past. This is an exercise applied by historians, economists, and sociologists, who inside the terms of "counterfactual" history investigate unprecedented configurations: "What would have happened, for example, and to go back to studies and positions concretely conducted and held in the past, if instead of investing in the rail network, there had been investment in the extension and improvement of the network of canals; if instead of building a vast system of highways, there had been a radical improvement of the diffused road network [...]"[34].

Robert Fogel, author in 1964 of *Railroads and American Economic Growth: Essays in Econometric History*, picks up on a tradition of hypothetical reconstruction of the past that gained a certain amount of renown in the 1930s, thanks to the anthology of texts published in England, *If It Had Happened Otherwise*, revised and reissued in those same years in the United States under the title *If: Or, History Rewritten*[35]. Famous figures and historians contributed to illustrate periods, personalities, or events, and constructed virtual histories in a narrative mode. Though not entirely rigorous, this attempt — which includes a piece by Winston Churchill — contains certain striking aspects. Imagining that certain things had not taken place, or that some of their consequences had differed from those effectively on record, these studies shed unexpected and critical light on any deterministic idea of history, on the concept of causality, on the unstable relationship between past, present, and future. Fogel, for example, tests the hypothesis of the central role of the railroad in the development of the United States, a hypothesis that has become axiomatic, and has therefore never been rigorously verified. What if instead of thousands of miles of rail, the same resources had been deployed to expand and build a dense network of canals?

A theory of change has to permit us to get out of hypotheses, scenarios, visions, and imaginaries that make boundaries – often too narrow – around the field of the possible, of the "legitimate". Therefore, it has to address both the past and the future. The project, inevitably conceived in the present, is positioned between the two extremes of past and future. Exercising its own faculties of hypothesis, the project can produce new knowledge about both.

[1] Berger, 1959, *Le monde en devenir*, *L'Encyclopédie Française*, t.XX: later published in BERGER 1964: 270.

[2] A copy of the manuscript is found in BERGER 1964.

[3] BERGER 1964: 116, "nous faisons l'expérience du devenir".

[4] KAHN 1962: 144.

[5] Also according to Polak, "Bergson's epistemology provides a new substructure for the continuing process of imagining the future", from the abridged edition of *The Image of the Future* (POLAK 1973: 143).

[6] "S'opère sans cesse une refonte radicale du tout" (BERGER 1964).

[7] *L'Encyclopédie Française*, t.XX: *Le monde en devenir*, 1959, later published in BERGER 1964. Fragment V bears the signature of Pierre Bertaux.

[8] Moreover, also in the *Dictionnaire Philosophique européen*, there are no entries for either *"avenir"* or *"futur"*, and in the *Enciclopedia Einaudi,* the theme of the future is reduced to simply the highly ideological debate of the 1970s.

[9] "Mais le sens que nous lui attribuons, soit explicitement dans nos discours, soit implicitement dans nos actes", Berger G. "L'idée d'avenir", in *Les Annales*, n° 118, 1960, later published in BERGER 1964: 227.

[10] In his project of *physique sociale,* as Comte renames the political sciences, the social system is fixed as the extension of the past and its linear continuation in the future.

[11] "Inversion du temps", Berger G. "L'idée d'avenir", in *Les Annales*, n° 118, 1960, later published in BERGER 1964: ° 236.

[12] "La disjonction, si douloureuse, qui s'opère pour nous, entre l'éternité dont nous rêvons et la durée qui semble nous englober et parfois nous étouffer. Elle nous enseignerait aussi, peut-être, que l'avenir de l'humanité ne rend pas dérisoire la destinée personnelle. Ainsi une phénoménologie de l'accélération pourrait-elle s'épanouir en une métaphysique de l'espérance" (BERGER 1964: 237).

[13] Berger, G., "La prospective", in *Revue des deux mondes*, 1957, February 1, later published in BERGER 1964: 219.

[14] "Vision en profondeur, qui cette fois, ne sera plus dans l'espace, mais dans le temps", *Encyclopédie Française*, t.XX: *Le monde en devenir*, 1959, later published in BERGER 1964: 216. Fragment V bears the signature of Pierre Bertaux.

[15] SECCHI 2005.

[16] BERGER 1964: 223.

[17] BERGER 1964: 224

[18] Berger, G., "L'attitude prospective", *Encyclopédie Française*, t.XX: *Le monde en devenir*, 1959, later published in BERGER 1964: 270-275.

[19] Benjamin, W. 1962, "Theses on the Philosophy of History", in *Angelus Novus - saggi e frammenti*, Torino, Einaudi (SAGGIO, 1940).

[20] LIPPITT, WATSON, WESTLEY 1958.

21 "Organizational concept", both in *The Year 2000* and in *Things to Come*. This means that the different conjectures are inserted in a socio-cultural, political, and economic mainstream that is made explicit in the construction of scenarios.

22 For example, Bertrand de Jouvenel pays tribute to him in the first pages of *L'art de la conjecture* (pp. 8), though he does not agree with the title of the lecture, which seems to see the future as something already existing and hence to be discovered.

23 For example Bernard Cazes, in HATEM 1993.

24 WELLS 1902 (from the edition 1913: 9).

25 One of the great trends recognized by Wells lies in the fact that "our dense populations are in a process of diffusion and areation", WELLS 1902 (from the edition 1913: 50).

26 WELLS 1902 (from the edition 1913: 30): a new, more critical way of looking at things.

27 WELLS 1902 (from the edition 1913: 32): "the possibility of an inductive future to correspond with the great inductive past of geology and archeology [...]".

28 "[...] les conditions de matérialisation et les conséquences d'hypothèses situées en dehors de ce que la majorité des experts tiennent pour le plus probable", Cazes, in HATEM 1993: 44.

29 KAHN, BRUCE-BRIGGS 1972, *Things to Come*, in particular the last section "The Ideology of Tomorrow".

30 In *Things to Come* Kahn clarifies his position: in the consultations, the aim is to clarify the actions to achieve a certain type of future, while in the texts for the general public, the position is more descriptive, though hardly without ideological prejudices.

31 "Pratique de l'hypothèse", LOI 2000: 116.

32 SECCHI, VIGANÒ 2004, with Fabian L., Costa A.

33 SECCHI, VIGANÒ 2004.

34 SECCHI, VIGANÒ 2004.

35 Squire J.C. (ed.), *If It Had Happened Otherwise*, Longmans, Green, 1931; *If: or, History Rewritten*, Viking, 1931.

The history of the territory of Modena reveals certain critical moments (which, as such, need not be separated with great chronological precision). In these moments the course of history changes its direction, leaving an older channel and taking a new one characterized by a different settlement structure and a different visible and social landscape that, in turn, constructs different problems, solutions, projects, and images. These are the periods in which the course of history of the city abandons its previous legitimacy to establish a new, different legitimacy. This is not an aspect related only to the history of Modena. Every city and every territory is marked by a more or less clearly recognizable set of paleo-channels of history.

THE MAP OF CARANDINI.

B. SECCHI, P. VIGANÒ, *SCENARI RETROATTIVI PER IL TERRITORIO DI MODENA: LA STORIA SI FA CON I "SE"*, IN *PER UN ATLANTE STORICO AMBIENTALE URBANO*, MAZZERI, C. (ED.), MODENA: EDIZIONI APM, SERIES "LE CITTÀ SOSTENIBILI", 2004 (RESEARCH AND TEXT IN COLLABORATION WITH: A. COSTA AND L. FABIAN. THE AUTHORS CITED IN THE FOLLOWING TEST HAV PREPARED SOME OF THE CHAPTERS OF THE ATLANTE).

The image, which comes from the research conducted on the Modena territory by Mario Panizza and others, seems particularly fertile. The paleochannel remains and deeply marks the territory; at a distance of years or even centuries, it is still recognizable. The reconstruction of the hydraulic regimen related to it and its possible evolutions is not a work of fantasy. The same signs that explain its abandonment often permit us to make valid hypotheses regarding how the territory would have been modified had the previous channel not been abandoned.

In the Modena territory, for example, examining only the most recent periods and without successive investigations, an initial turning point is located in the Napoleonic period. Its results are well represented in the "Carta Topografica del Ducato di Modena", which is the outcome of surveys made from 1821 to 1828 by the colonel Giuseppe Carandini.

A change in the system of land ownership revealed by the research of Marco Cattini has the consequence, in this period, of related agricultural development, which had been considered quite backwards by all external observers up to that point, alongside the growth of the "diffused rural city", a scattered rural settlement with a density of about 12 farmhouses per km². The construction of the landscape of the "*piantata*" and its architecture, well described by Cattini, is the most visible aspect of this.

"Each field of rectangular form, on its two long sides, aligned festoons of vineshoots wed with trees, placed at intervals of about one meter from each other. Like a grand natural architecture, the mixed cultivation of the 'piantata' was organized on three distinct, overlapping levels: at ground level, the cereals, corn, legumes and fodder; halfway up, along the two larger sides of each field, the hanging vines whose leaves, intertwining, spread from one trunk to the next; finally, above the festoons of vines, there exploded the flourishing foliage of the elms, pruned to multiply the younger branches and to obtain each year the greatest possible abundance of leaves". In the whole Po Valley area, the landscape of the "piantata" had a quality and a density on par – though not similar – with the most famous landscapes of central Italy. Halfway through the 19th century, the diffused rural city, in its demographic dimensions, was equal to the city proper. As in other Italian and European regions, the spread of settlement in rural areas is an ancient phenomenon.

A second turning point comes in the 1930s: the growth of the urban outskirts is evident at this point, while that of the diffused rural city stops. The landscape of the "piantata" begins to vanish until it is almost totally destroyed in the 1950s,

THE MAP OF CARANDINI AND THE TOPOGRAPHICAL MAPS OF THE IGM OF 1881-1893 AND 1935 DEMONSTRATE A SIGNIFICANT DISPERSION OF COUNTRY HOUSES IN THE TERRITORY OF MODENA. WHAT EMERGES IS THE MODEL OF THE "PIANTATA" WITH THE AVERAGE SIZE OF THE TYPICAL SHARECROPPED FARM IN THE MODENA AREA AND THE ASSOCIATED FARMHOUSES, IN DISPERSION THAT IMPLIES A DENSITY OF 12 COUNTRY HOUSES PER km² (WITH AN AVERAGE AREA OF 8.33 HECTARES).

REGGIO EMILIA IN THE FIRST HALF
OF THE 18TH CENTURY, DRAWING
BY WERNER OF THE LANDSCAPE
OF THE PO VALLEY "PIANTATA", IN
PELLINI, U. 1996, ALBERI NELLA
STORIA DI REGGIO.

and not only in the Modena area. At the same time and as a side effect, the biopotential of the territory studied by Boraldi decreases until it is seriously compromised in the 1970s. Industrialization and urbanization, a lack of innovation in the agricultural sector, and the rise in the period between the two world wars of extensive crops, lead to modification of the Modena and Po Valley landscape. This is a modification generally perceived as decay.

Problems connected with sustainability are not yet perceived as important by the shared sensibility, and have little impact on the direction taken by urban growth. Other questions are the priorities of the collective imaginary and urban planning policy: those of the urban form, seen as correspondence between the organization of the city and that of the territory, those of inhabitable and productive space, seen as the correct construction of spatial relations between different activities and different building types, those of maximum density of edification, seen as a tool to check the rise of real estate values, on the one hand, and to improve general hygiene conditions, on the other.

The "waterproofing" of the territory, however, becomes an important phenomenon in this period. The traditional hydraulic network that, to a great extent, relied on canals inside the urban settlement itself is replaced, despite the late construction of the urban sewer system, but in a progressive engineering of the territory, by a network that relies widely on works of rectification of waterways and their transformation into drainage canals for the removal of water arriving from the urban sewer system.

A third turning point, finally, comes at the end of the 1970s and in the early 1980s when formation of a diffuse city of urban origin begins, with the great expansion of

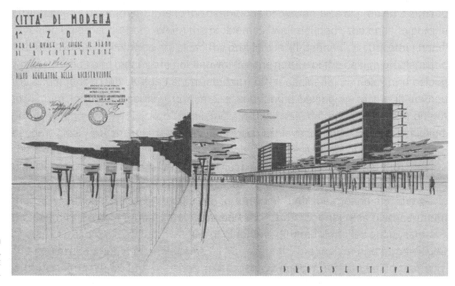

M. PUCCI, RECONSTRUCTION PLAN
OF MODENA, 1947, PERSPECTIVE,
ARCHIVIO STORICO DEL COMUNE
DI MODENA.

urbanized areas and relative infrastructures, the diminishing of biopotential, the increasingly radical waterproofing of the territory, the reduction in density of the rural and urban landscapes.

Trend scenarios

As is obvious – and shown by the research conducted by the different groups – these turning points and intermediate periods are closely connected to profound changes in the social structure of the city and territory of Modena, to a profound change in the relationships between the city and the rural population, to changes in their economies, in the behaviors of the different social groups and, finally, in the formal and informal structure of power. But where the city and the territory are concerned, each of these breakthroughs is probably not the inevitable consequence of such changes and modifications. Other individual and collective choices, in the various fields, could have been made had the past been interpreted differently, had the present been differently understood and the future differently imagined.

First of all, this implies the possibility of outlining certain trend scenarios: what would have happened had the trends of each period continued without interruption or sudden breaks? What conditions would have been necessary to permit continuity of their essential lines, even in the altered social, economic, and cultural structure? The point is obviously relevant where sustainability is concerned. Short exercises can clarify several themes. For example:

First scenario: settlement dispersion.   The map of Carandini and the topographical maps of the IGM (Istituto Geografico Militare Italiano) of 1881-1893 and 1935, revised by the Institute for the Artistic, Cultural, and Natural Resources of the Emilia-Romagna Region, demonstrate a significant dispersion of farmhouses in the territory of Modena (Pezzoli); a dispersion that implies a density of inhabitants and farmhouses (8,33 inhab/km^2) greater than that of neighboring areas, and that stops in 1935 when the growth of the rural population ceases (Cattini). What would have happened had this settlement mode continued beyond 1935, even in the case of greater demographic growth? What would the situation be, as compared to what has effectively happened, in terms of dispersion, after 1980? Or, in other words, if the modes of growth of Modena had always been those of a city balanced in its relationship between population in the rural territory and population in the city?

LAYERED REPRESENTATION OF LAND OCCUPATION IN THE CITY OF MODENA, BASED ON DATA COMPILED BY THE IBC USING THE CARANDINI MAP AND THE MAPS OF IGM, 1881-1893.

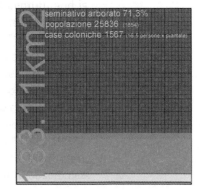

Second scenario: forms of industrial growth in.   Modena, as in many other Italian cities, have followed a model of industrial growth that is centralized and in contiguity with the compact city, giving rise to the birth of the modern periphery. More or less clearly defined industrial zones are put into relation with major infrastructures at the edge of the compact city.

CALCULATION OF THE NUMBER
OF COUNTRY HOUSES IF THE
ENTIRE MUNICIPAL TERRITORY
SUITABLE FOR SETTLEMENT
(96,12%) HAD BEEN URBANIZED
IN KEEPING WITH THE MODEL OF
THE "PIANTATA".

In other European situations, in Flanders or Veneto, for example, the model of industrial growth has been different, and has always been that of dispersed organization. What would have happened had Modena and Emilia chosen the Venetian-Flemish model? What would have been the results for the landscape? And for biopotential?

Actually, by extending and integrating the cognitive and analytical elements to the vast super-municipal area, we can better interpret the complex processes of expansion of the minor centers. The towns of the Modena area have, in fact, partially followed the same path as the city, keeping the historical center as the central point of reference while at the same time permitting the fragmentation of the settlements in multiple points of a territory until they are only slightly occupied, or left without settlement. The combined impact of the two tendencies determines an urban continuum comparable to the most classic examples of dispersion.

Many of the causes of these different tendencies can be found in the different temporal unfolding of industrial development in the areas taken into consideration.

Had more been known: the role of information in territorial policies.   Every choice is made in imperfect conditions of knowledge. An idea that runs through the construction of information systems is that by increasing accessibility of available data, by "systematically" organizing them, the systems can contribute to improving decision-making processes. This leads us to wonder what would have happened in the past had it been possible to obtain at least the information we have today, especially regarding issues of environmental sustainability: quality of soil, water, air, natural biopotential.

CALCULATION OF THE NUMBER OF
POTENTIAL INHABITANTS IF THE
ENTIRE MUNICIPAL TERRITORY
SUITABLE FOR SETTLEMENT
(96,12%) HAD BEEN URBANIZED
IN KEEPING WITH THE MODEL OF
THE "PIANTATA".

Obviously the supply of adequate instrumentation is itself the result of desire and awareness, of sensitivity to themes that were utterly overlooked in the past. The observation of maps, with their informative content, demonstrates that human beings see what they want to see. From a certain point on, until recently, settlement dispersion is not recorded as a specific phenomenon.

Information/images

Every decision-making process is strongly influenced by certain essential suppositions that are often represented in images, which nevertheless can lead to reductions or simplifications of complexity. The set of these images constructs a collective and individual imaginary, which we often can recognize as being deposited in certain projects that have marked the history of the city and the territory.

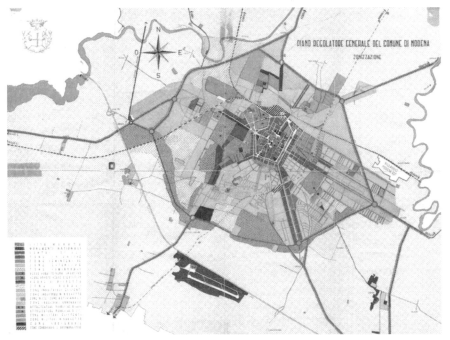

G. CAMPOS VENUTI, O.
PIACENTINI, L. AIRALDI, C.
MORELLI, A. TOSI, MUNICIPAL
TECHNICAL OFFICE, MASTER
PLAN OF THE CITY OF MODENA,
1965, UFFICIO STUDI E ARCHIVIO
DI DEPOSITO DEL COMUNE DI
MODENA.

Once again, there are very simple examples. The image conveyed to us by the projects of 1958 of Mario Pucci for the Master Plan – never approved – and of Giuseppe Campos Venuti for the Master Plan of 1965, form a contrast. One is rooted in a long history, attempting to prolong it, and constructs the idea of a city that is distinguished from the surrounding territory; in a certain sense it pursues

M. PUCCI WITH THE PUBLIC
WORKS OFFICE OF THE CITY OF
MODENA, MASTER PLAN OF THE
CITY OF MODENA, ZONING, 1958,
ARCHIVIO DI DEPOSITO DEL
COMUNE DI MODENA.

the figure of the *finitudo*. The other, instead, opens to the Po Valley territory, redesigning it as a whole, and constructs the idea of a territorial policy that is able to get beyond the unsatisfactory results of past history.

Would these images and their related design proposals have been different had it been possible for their authors, and for the context in which they worked and discussed their ideas, to have more detailed, extensive information? What would have needed to be taken into account at the time by the subjects who took part in the making of certain important decisions? How would the processes have developed that have slowly led to the realization of those images, their modification or distortion, or finally their replacement by other images?

It is useful to retrace this history and the forks in the path that could have emerged. The idea is to talk about the future more than the past.

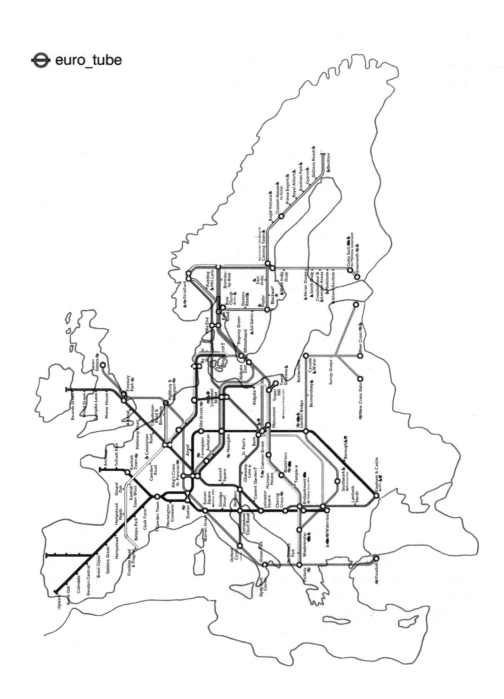

euro_tube

The image of the future

The construction of an image of the future[1] is at the center of the thinking of Fred Polak, who, in the same years as Gaston Berger, combines reflection on time with reflection on space. The invention of a world, of another space, proceeds alongside that of time. Images of the future and spatial concepts are closely interconnected. Polak's thesis is that after the experience of World War II, and for the first time in the western world, the capacity to think about the future was drastically reduced, together with a shrinking of the temporary horizon, generating an incapacity for "self-correction" and for renewal in time. His text *Die Toekomst Is Verleden Tijd (The Image of the Future)*, published in two volumes in 1955, centers – like the writings of Berger – on the construction of a new man. In the 1973 preface to the translation and synthesis in English, Elise Boulding talks about the strength of spirit of a Jew who lived in hiding during the war years in occupied Holland. From that experience of destruction, he absorbed "a vision of man as a future-creating being", harnessing "his own intellectual and spiritual energies to communicate this vision to precisely that generation of young men and women who need desperately to know that it is possible to create other and better futures"[2].

Analysis of collective and public images of the future, even prior to the production of new images, seems to Polak to be the decisive activity for a new interdisciplinary social science that sets out to orient change. 'The whole question of public policy in relation to image-change may be a matter of life and death for Western civilization', writes Polak, so "we cannot accept the idea of controlled image-change as a social good per se"[3].

The power of images of the future – developed inside the educational and political system to help focus choices and actions, as well as in the definitive orientation and, in some cases, construction of the future – is well known. Polak's concerns are more pertinent than ever today: "The rise and fall of images of the future precedes or accompanies the rise and fall of cultures"[4]. For this reason, the "image of the future" simultaneously represents to Polak the object of the research, the statement of the problem, and its manifesto[5].

Amidst changes in the social structure and ideas of future, what are the relationships between images of the future and the future itself? While Polak's text

NEW TERRITORIES: SCENARIOS FOR THE EUROPEAN CITY. "EURO_TUBE", FROM VIGANÒ, P., ED., 2004, NEW TERRITORIES.

reduces the field of his personal research to the analysis of utopian images, sifting through the past of western culture and proceeding from it to the contemporary age, his study is primarily interesting for the clear juxtaposition of the two terms *future* and *image*. It is only through images that we can represent what is to come, and these images developed each time in a specific present speak to us of both.

Bernard Cazes (1986) also retraces different ways of representing the future, but adds that the "elementary forms of representation of what is to come" are few and can be linked back to a repertory of "figures of the future" organized around the fundamental concepts of western culture: modernity and progress[6]. According to the author, the association "M+P+", the extension of the concepts of modernity and progress and their reinforcement as guiding values of the society, is typical of a positivist, evolutionist vision challenged by "uchronias" (extending the notion of utopia, connected with space, into the dimension of time). The expression M-P+, on the other hand, covers all the figures and images of retrogression and progress "*par antimodernité*"; M+P- covers anti-utopias; and, finally M-P- covers the images and figures of a future of decadence. While apparently simple and limited, the four sectors proposed by Cazes make it possible to position a great many efforts to imagine the future, from traditional science fiction all the way to scenarios for the city and the territory.

Scenarios and images

Heavy trends.   Traffic in Towns, the Buchanan Report published in 1963, takes the year 2010 as its temporal horizon. It is one of the best-known exercises in the construction of scenarios and images of the city of the future in relation to the growing rate of motor vehicle usage. Starting from a "heavy trend", as Jouvenel would call it, or one that would be very difficult to reverse, such as the race towards individual car ownership and use, the group led by Buchanan analyzes in depth the capacity of urban spaces to absorb the new quantities of vehicles, and explores the adjustments that may have to be made in the existing fabric and the newly conceived city.

SKETCHES OF THE TWO
DIAGRAMS CONTAINED IN
TRAFFIC IN TOWNS, 1963:
DIAGRAM OF CIRCULATION IN
A HOSPITAL THE CELLULAR
CONCEPT.

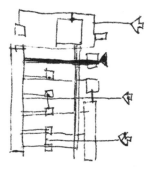
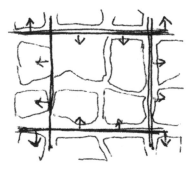

The proposed spatial devices are the result of the conceptual reformulation of the problem of traffic and the theme of urban mobility. The metaphor and diagram of the hospital used by Buchanan, with the separation of flows and the isolation of rooms ("environmental units"), suggests the construction of multiple stacked levels, new street patterns, and sectors to keep buildings from facing thoroughfares. In definitive terms, the public space of the street, which in its traditional configuration is not able to solve the problem of the presence of the automobile, is rethought within a new conceptualization of movement and therefore the use of space.

Buchanan's long-term scenario relies, according to Jouvenel, on an "implicit prediction" that takes form, for example, around the offering of a new mass technology to society, and the general desire to make use of it. What is interesting about the report is that it isolates a strong trend and makes the rest – the entire transformation and conception of the city – revolve around it. The scenario analyzes the consequences for the urban landscape of congestion, and the proposed conceptual passage leads to a new city image. Just as a hospital regulates its flows by diversifying and separating spaces (rooms, halls, and corridors), so too can the conflicts caused by traffic be resolved or reduced in a city conceived as a composition of "urban rooms" and "urban corridors"[7]. In the "cellular concept",

depicted in a diagram without scale taken from the "precincts" proposed by Alker Tripp in the 1940s (inserted in the lengthier reflection on the cellular metaphor I outline in the first part of this book), multiple cells are bordered by circulation streets that separate them from the others; the urban rooms are not cross-passages, and the nature of the corridor is to punctually serve them. These are unlike the sectors of Le Corbusier, which have a smaller size, but distant city parts, each separated by green spaces and contained inside a circulation grid of a higher order.

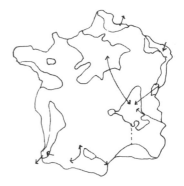

The exploratory nature of the scenario appears in the "Practical studies" of *Traffic in Towns*, which are not – despite the title – applications of the theories formulated[8], but are rather experiments in which the theory is itself developed inside concrete situations simulating all the conditions of the design experiment. The representation of urban space, on different scales and using multiple modes (diagrams, schemata, maps, perspective sketches), plays a central role in the Buchanan Report and sustained the rapid spread of the theory of sectors, one of the (few) victorious theories of modern urban planning[9].

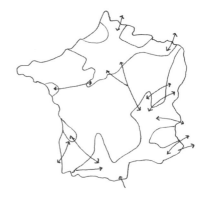

Image of the unacceptable.   A few years later another "image" has a strong influence, and not only inside the debate on the scenario. It is the "scenario of the unacceptable" contained in *Une image de la France en l'an 2000*, research published in a reduced form in 1971, and part of the preparatory work on the *Schéma général d'aménagement de la France*, commissioned by the DATAR[10]. The image springs from a tendential scenario, or one that takes the trends perceptible at the moment of its construction as its backdrop. Having abandoned the first hypothesis of selecting an image "of contrast" and explaining the path required to reach it — a route that was not feasible due to too many inventoriable questions and the lack of reliance on sociological knowledge on the part of the French — the scenario starts with data from the present to move towards the future, extrapolating trends and excluding the most radical eventualities, such as a new revolution. The resulting image, according to the authors, has three qualities. It is an element of reference that permits the situating of other phenomena and policies. It is a *"repoussoir"* image, repulsive and contrasting, useful for mobilizing reactions in relation to the progress of phenomena in the present. Finally, it is an image that, by displaying its limits of conception, opens up ulterior degrees of freedom of action. The junction points and the forks along the way appear where the image is blurred[11].

After the images, *"la France de cent millions d'habitants"*, *"la France côtière"*, *"l'agriculture sans terre"* ("'France of one hundred million inhabitants'", "'coastal France'", "'agriculture without land'")[12], that introduce the DATAR to the prospective reflections, the image of France in the year 2000 is developed from the evolution of four types of societies: urban, industrial, agricultural, and rural. In 2000, France is imagined as being increasingly urban and yet more polarized to the detriment of areas that are depressed and excluded from the flows; the state has lost many of its prerogatives, but the new regional power cannot manage to limit the aggressive operation of the large international corporations; in reaction to this situation, there is a return to nationalism… The prospects are gloomy, though the picture offers much food for thought. It is the "unacceptable" scenario that indicates lines of research and demands the taking of positions. Often the function of the scenario is precisely to show what we do *not* want, to shed light on a future that is *not* part of our desires. Herein lies its forceful constructive role for the future, in opposition to what might happen.

Beyond the Blue Banana?   Some important scenarios for the city and the territory address fields that architects and urbanists cannot approach on their own. Many scenarios involve a wider-ranging discussion including other forms of knowledge and disciplines or different social groups involved in the project. Even the most vast scenarios, however, have to be tested, and they require a territory-laboratory if their consequences are to be understood. The construction of scenarios solicits a dual approach: of great abstraction and, at the same time, minutely descriptive, paying attention to contingent factors. Scenarios can be the starting point of a discussion, not its end.

Map-data-scenarios.   Maps spatialize phenomena. They generally come already prepared from different archives, and they are a basic tool in the construction of scenarios. They are an intermediate layer between specific situations and scenarios (situations – maps – scenarios). Maps are already a representation of trends. They approach the scenario that, in its simplest expression, relies on them. On the other hand, trends can be radicalized so that they can express all their potential, or the risks they contain.

What trends and what scenarios can be represented on the European scale? Looking at a number of attempts made by geographers, architects, planners, or artists over the last few decades, one recognizes a range of approaches.

The first attempt sets out to specialize the European territory, or imagines recognition of specializations in the various regions or parts of Europe. In general, the idea is to break up the continuity of physical space into separate elements, parts that can be interpreted in keeping with models of integration or separation, demonstrating role changes that have happened or could happen, or the different rhythms of development of European regions. The image of the *Blue Banana*, though sharply criticized for its simplification and reduction of European complexity, has become a powerful tool of territorial marketing, a stabilizer of differences between the various parts. Proposed by Roger Brunet in 1989[13], the Blue Banana distinguishes Europe in a rhetorical operation of great impact, between active and passive zones, thus legitimizing ambitions that have not always been realized. Not everything that lies inside the "banana", in fact, takes part in the dynamism and the density of activity that is supposed to set the zone apart from the rest. For example, Wallonia, one of the crucial places of the European industrial revolution, which has been going through a deep structural crisis for several decades, is part of the blue zone. However, its presence there may only be a matter of wishful thinking as it struggles to move towards a new type of development, of transformation of its enormous legacy of infrastructures and areas. The image of the Blue Banana has produced – as is only logical – its opposite, the so-called 'Sunbelt' region along the Mediterranean coasts, where the good climate attracts new businesses, as in the American Sun Belt, or simply new tourists (a reading that is closer to reality and to trends).

Many scenarios can be represented as forms of discretization of the European territory: a Europe of megacities, a Europe settled in increasingly dispersed ways, but at the same time with new concentrations of diffused urbanness; an elderly Europe composed of "old age homes", regions that have made the presence and concentration of a population of senior citizens into a new vehicle of growth; *park territories* recognizable thanks to their extraordinary landscapes, often the result of lengthy sedimentation; and so on. New geographies are radically rewriting the European territory, changing the role of its regions, and redefining its future image. The idea that Europe is made of differences, that these differences have a geographical basis, and that a strategy for the future can or should reinforce

these differences remains a strong, resistant point, both in the local and in the global debate. The European territory is interpreted, within this first position, as a *territory of differences*. Its future is imagined to be that of a place of ripening and growth, free and democratic – of differences.

The second approach is the opposite of the first: the accent is on things places have in common, the pervasiveness of certain phenomena, and the possibility that they can have an impact and be relevant for the entire continent or even the whole planet. Starting with population density and trends in progress, in the 1960s Doxiadis imagines Ecumenopolis, a covering proximity pattern that determines a single urban condition. Doxiadis' scenario is confirmed today by the statistics of the United Nations which indicate that in the developed nations of the West, 80% of the population lives in urban settlements; while in the south of the world, the percentage is 45%, slated to increase to 60% towards 2025[14]. The continuous chain caused by increasing population density is indicated as a new state of urbanization that forces us to change our preconceived ideas about its organization. In this case, the territory is the inert surface for the driving forces of urbanization that bring homogeneity to situations that were previously marked by discontinuity.

This second family of map-scenarios contains a version that carves out its own position: here the continuity of European ground is a new spatial organization, separate from the existing one. The drawings of Constant for New Babylon, made from the 1950s to the start of the 1970s, are among the strongest representations of this scenario. The new form of urbanization is at the same time a territorial infrastructure and a support for new lifestyles. While the extreme version of this position is New Babylon, some of the contemporary images of the European structural network, for example that of high-speed trains, seem to move in the same direction. Europe can be represented in its present and, above all, in its future conditions as an extended, continuous infrastructural space of tubes and tunnels that are capsularized (DE CAUTER 2005), almost entirely detached from specific geographical situations. Overlaying the existence of a new continuous space would guarantee greater freedom and uniform access to a wider field of possibilities, especially linked to individual processes of organization. This intention, however, also produces new polarizations and marginal situations, or reinforces old ones. The theme of continuity and homogeneity of access to inhabited space and the question of the infrastructural offering across Europe, as conditions that are opposed to localism and specificity, touch on crucial aspects of Europe's future.

The scenarios of interest pertain to the characteristics of space and the form of the territory, introduced in reflections about the future. For its logical and illustra-

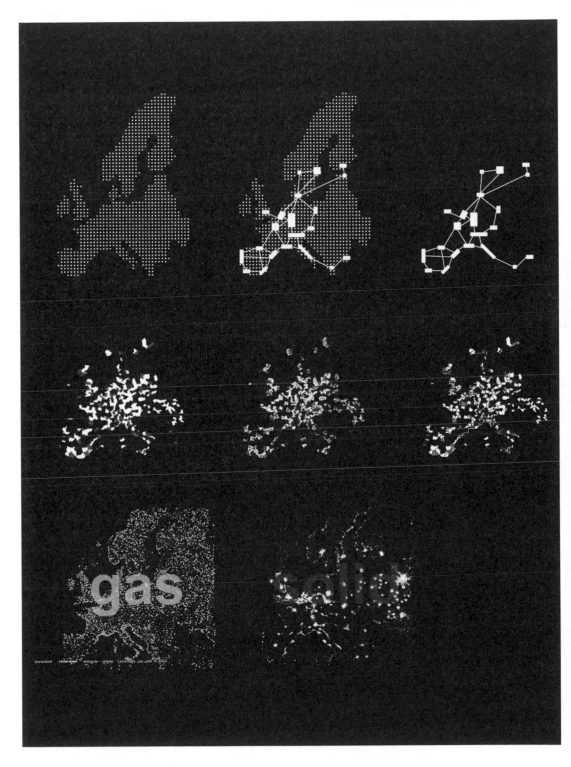

tive clarity, it is worth recalling an exercise from the mid-1990s that contributed to putting the construction of scenarios back into the practice of urban and territorial design. The opportunity came from consideration of the large modern periphery and a concrete episode, a workshop done in the early 1990s on the future of the Alexander Polder and, in particular, the development by Van den Broek & Bakema in Rotterdam, built in the 1950s, that is one of the most important works of Dutch modern and structuralist urbanism in addition to being the expression of an urban planning "mean" that makes it representative of an era. Precisely its normality permits Rem Koolhaas to approach the theme – thenceforth at the center of the urban planning debate in Holland – of the transformation of the modern city on different scales: from the urban scale to the use of different portions of the national territory. The two well known scenarios/models, of the Point City and the South City, highlight some of the settlement trends towards the concentration of activities in the south of Holland, on one hand, or in the green heart of the Randstad, on the other. As opposed to clichés like the one that still imagines an intact green heart in the ring-city, the two scenarios take certain trends to paradoxical extremes, applying a powerful rhetorical tool: the whole Dutch population is imaged as being concentrated in one point, while the rest of the territory becomes a great green park with monuments; or all the settlements move south, leaving

SKETCH OF THE TWO SCENARIOS
POINT CITY AND SOUTH CITY
PROPOSED BY REM KOOLHAAS IN
THE STUDY ON THE ALEXANDER
POLDER, 1993.

the north and Frisia to a future as a recreation park, with a marginal status. The hypotheses cross scales from nation to neighborhood to street, as well as ways of living, and demonstrate that when we talk about density, nothing has yet been said about the characteristics of space, that the density of the modern district, its "normality"[15], is made of assumptions that can be overturned. It proves that different models do exist.

Reworking data, maps and diagrams, prospective sketches, comparisons: everything is expressed in synthetic and simultaneously effective ways. In the short text that accompanies the exercise, Rem Koolhaas retraces the last decades of Dutch urban development, of erosion and caricature of the "green heart", of urban planning and political discourses entirely detached from concrete transformations. He urges less ambiguous positions. The two scenarios for Holland based on extreme density have the value – according to Koolhaas – of reintroducing "explicit ideological choices" (KOOLHAAS 1993a), of putting different ideas about the future at the center of the discussion.

The above examples demonstrate that situations and trends are subjected, in the scenario, to a process of spatialization in which the third dimension plays the fundamental role of putting the hypotheses, even radical ones, in relation to the phys-

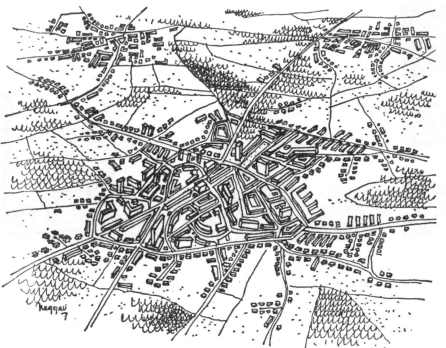

SCENARIOS/VISIONS TAKEN
FROM GUTKIND, E.A. 1962, THE
TWILIGHT OF CITIES:

ANTAGONISM OF TOWN AND
COUNTRY.
LANDSCAPE TOLERATED.
IMPERFECT DECENTRALIZATION.
HIERARCHY OF LOCALITIES.
THE MONSTER WITH TENTACLES.
SOCIAL STRATIFICATION.
CULTURAL CONCENTRATION.
INDUSTRIAL DISORDER.
RURAL ISOLATION.
ARCHAIC FARMING UNITS.
PRODUCTION COMMANDS.

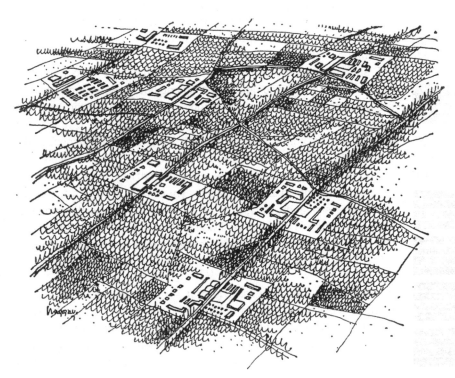

ONE LIVING AREA.
LANDSCAPE OMNIPRESENT.
PERFECT DISPERSAL.
EQUALITY OF COMMUNITIES.
THE LOOSE CLUSTER.
SOCIAL AFFINITY.
CULTURAL UBIQUITOUSNESS.
INDUSTRIAL ISOLATION.
RURAL INTEGRATION.
RATIONAL FARMING UNITS.
CONSUMPTION DEMANDS.

ical environment that they impact – its subjects and its users. Scenarios should be imagined within a process of transformation, defined with comparable levels of abstraction.

The texts that accompany the scenario often have the character of statements, assertions, and hypotheses that clarify the addressed theme. They are not manifestos but logical constructs that test the applicable hypotheses. They are narrative sequences of plausible events, telling stories and illuminating dynamics. The purpose is not to overwhelm a place with new forms and structures, but to impact the territory of possibility.

An important part of the thinking about tools addresses the relationship between forecast and scenario, between processing data and hypothetical reflection. Developing scenarios calls for quantitative groundwork, which is a labor of simple extrapolation of trends, as Godet writes, picking up on the lesson of Berger. It is equally fundamental to explore a prospective approach in order to render explicit and articulate the underlying hypotheses behind quantitative forecasts, which are not always positioned on a horizon of continuity (GODET 1992).

The scenario makes it possible in a less conflicted setting to raise difficult questions about which it is not possible to speak. It permits the testing of concepts, strategies, priorities, and innovations. It stimulates reactions, integrates other approaches, and widens the debate[16]. It generates images outside the routine.

[1] Fred L. Polak, *Die Toekomst Is Verleden Tijd*, concluded in 1953 and published in two volumes in 1955, then translated into English and published in 1961, again in two volumes, under the title *The Image of the Future*; finally abridged to a single volume in 1973 for the Elsevier publishing house (Amsterdam, London, New York) by the same translator, Elise Boulding. She was the wife of the English economist who emigrated to the United States, Kenneth Ewert Boulding, Polak's neighbor during the year he spent at the Stanford Research Institute (1954). I refer to this later version, from 1973, which notably gives the following title to the first chapter: "The Future as a Work of Reconstruction".

[2] Boulding, E., introduction to *The Image of the Future*, 1973.

[3] POLAK 1973: 15.

[4] POLAK 1973: 19.

[5] "We are treating the image of the future both as an object for research and as a statement of the problem" (POLAK 1973: 22).

[6] CAZES 1986: 29-30, chap. 1, *Continuité et mutations dans l'histoire de la pensée prospective*.

[7] *Traffic in Towns, A study of long term problems of traffic in urban areas*. Reports of the Steering Group and Working Group appointed by the Minister of Transport, London: Her Majesty's Stationary Office, 1963, pp. 42.

[8] *Traffic in Towns*: "...we arrived at our working theory by a combined process of searching for principles and studying actual cases", pp. 53.

[9] MANGIN 2004.

[10] OTAM 1971a.

[11] Jacques Laigroz, 1971, introduction to OTAM, *Une image de la France en l'an 2000*.

[12] OTAM 1971b.

13 See a recent comment in BRUNET 2002.
14 In the 10th Venice Architecture Biennale in 2006, Richard Burdett takes his cue precisely from this new urban condition.
15 LOMBARDO 2008; SECCHI 2008.
16 SMITH 2007: 85; BOZZUTO, COSTA, FABIAN, PELLEGRINI 2008.

The context in which the program of new social housing in Brussels is inserted is one of great environmental quality. The presence of agricultural activities and the proximity to the forest and to a wetlands zone are the important qualitative features. Next to the site are calm residential areas with different and clearly distinguished spatial and social characteristics. Therefore, deciding on and justifying the density and number of housing units to build take on important roles in the project, and, in our view, must be evaluated starting with the physical and social conditions of a larger area: on the metropolitan and local scale.

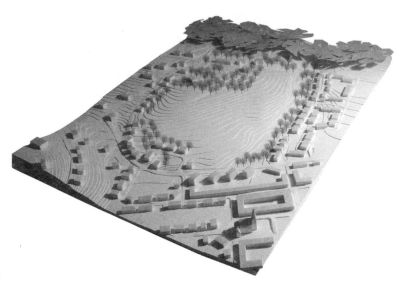

MODEL OF THE SITE.

B. SECCHI, P. VIGANÒ, COMPETITION FOR A NEW
PUBLIC HOUSING DEVELOPMENT IN BRUSSELS,
2008, WITH A. CALÒ, D. CHANG, S. DUSSIN, G. FINI,
S. PELUSO, IN COLLABORATION WITH CITY TOOLS
(N. HEMERLESS, A. CRAHAY), LBL ARCHITECTS
(G. LELOUTRE, V. BRUNFAUT).

We have avoided the facile solution of presenting scenarios of density variation – a high, medium, or low density scenario, a move that can suggest the often erroneous idea that *in medio stat virtus*. Instead, our interpretation is that the nature and characteristics of the physical and social context suggest a roughly optimal density capable of providing a responsible and economically feasible response to the social housing deficit in Brussels. This density cannot be exceeded without seriously compromising the future environmental and social qualities of the site. This allows for modest upward or downward variations, defining an order of magnitude, not inflexible limits.

Therefore, the proposed scenarios imply similar densities, while the forms of inhabitable space and the character of the public spaces vary from one interpretation to the other.

ICON HALEN
HERRENSCHWANDEN, BERN,
ATELIER 5, 1955-1961;

ICON, HOUSES AT KAUTTUA,
ALVAR AALTO, 1939;

The theme at the center of the project has to do with *"comment vivre ensemble"*, and the scenarios can structure an in-depth discussion on these aspects. How

ICON, WEST RESIDENTIAL UNIT,
IVREA, ITALY, GABETTI E ISOLA
ARCHITETTI, 1969-1971.

can we insert, amidst existing parts of the city, a new part that continues the construction of the urban collage, offering new spaces for individual and collective life?

Neither imitative nor a manifesto, the new city section has to justify itself through the ways of living that it promotes and through its qualities; through the relationships it generates between public and private space, between the city and the forest, between perception of density and measurable density. The three scenarios proposed are the expression of this position.

Ceci n'est pas un projet

The role of the scenarios is different: they are not limited to offering alternative solutions from which to choose, but instead urge evaluation of the consequences of each proposal, to clarify the goals we want to and can achieve in a given physical and social context. The development of scenarios is not an abstract, absolute process. It is based on careful interpretation of the context in which we recognize hard constraints – connected with the terrain, its topography – and more malleable aspects connected with the urban morphology, social relations, and the general perception of the territory.

VIEW OF THE SCENARIO OF
INHABITED FABRIC
WHICH CONSTRUCTS DENSE
EDIFICATION IN CONTINUITY WITH
THE CHARACTERISTIC FABRIC OF
THIS PART OF BRUSSELS.
ABOVE.

VIEW OF THE SETTLEMENT FROM
THE EXISTING NEIGHBORHOODS
AND TOWARDS THE FOREST;
BELOW: VIEW OF THE NEW FRONT
FROM THE LOW POINT OF WATER
GATHERING.

The topography is that of an exceptional landscape. The pastoral dimension of the site opens the imagination and represents an entire region: Brabant and its landscape of alternating forests, inhabited settlements and cultivated fields. Towards the city, a large part of the second belt of Brussels appears as a collage of neighborhoods with very different and recognizable morphologies. The east of Woluwe-St-Pierre is exemplary in this sense, and the Dames Blanches site is located at the hinge of multiple pieces of the puzzle, between garden city, modern developments and large villas with swimming pools, corresponding to different urban conceptions.

In the three scenarios the hard constraints will become constants. Among them, the utilization of the compluvium: the green belt of terrain between the forest of Soignes and Rue des Dames Blanches will be used to collect the water from the waterproofed surfaces of the new neighborhood.

Comment vivre ensemble: three scenarios

The three scenarios proposed do not differ in terms of density, quantity, or environmental quality, nor in the envisioned equipment, cost, or proportion of social housing (65%).

The differences pertain to *"comment vivre ensemble"*, meaning the ways of living they offer and the relationships they weave with the existing neighborhoods, thereby developing different urban sequences of transition and exploring distinct ideas of urbanness.

What brings about difference may cause reactions on the part of the actors and subjects involved: the regional authorities, local administrators, inhabitants of nearby zones… The purpose of the three scenarios is to define a grid of possibilities and reasoning that can help organize the debate and the negotiations.

Real density.   At first glance, real density seems like the fundamental variable of the project and the debate that will take place around this report. It is the fundamental variable for certain persons involved, such as the politicians who have set the number of housing units to be created inside the Brussels area, and have therefore, by extension, set objectives of density.

It is worth pointing out that no strong and univocal link exists between density, building types, and the social level of their inhabitants, as demonstrated by our interpretation of the city parts. If the densities called for at times

VIEW OF THE *INHABITED STRIPS* SCENARIO WHICH PERMITS CONTINUITY OF OPEN SPACE, OPENING THE BUILDINGS TO THE LANDSCAPE THANKS TO LARGE "WINDOWS", FOLLOWING THE CONTOURS OF THE TERRAIN.

ABOVE: VIEW FROM ABOVE, OF THE HILL TOWARDS THE CITY;

BELOW: VIEW OF THE EXISTING NEIGHBORHOOD TOWARDS THE FOREST.

by the Brussels Region reach 100 housing units per hectare, densities we find in certain "innovative" neighborhoods in Europe, greater precision is nevertheless required. The Borneo Sporenburg district in Amsterdam, for example, indicates a real density, excluding public spaces, of 100 units/hectare, while Leidsche Rijn in Utrecht has an overall density (public spaces included) of 37 units/hectare.

We could therefore compare the three scenarios to European references regarding overall density (the ratio between the number of housing units and the area of the site), and real density (the ratio between the number of housing units and the area without public spaces and zones where edification cannot take place). It is from this latter vantage point that the three scenarios differ: they propose very different types and sizes of public spaces.

Perceived density.   All the research on the theme of density, which has been substantial in recent years, underlines the important *décalage* between "real density" and "perceived density", above all from the perspective of the inhabitants who await and fear the arrival of the project, such as those who live around the Rue des Dames Blanches. For a large portion of the people and figures involved, density expressed in terms of quantitative relationships will not be the most important variable in the debate, simply because it cannot be easily calculated relying on

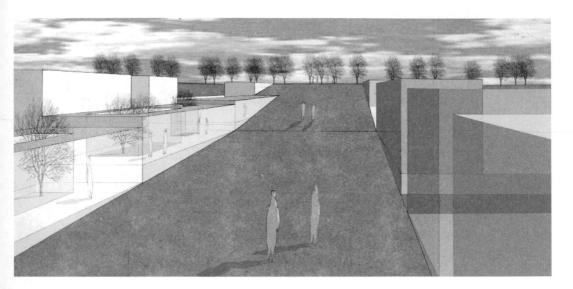

VIEWS OF THE *INHABITED LEAF*
SCENARIO WHICH INSERTS
COMPACT CONSTRUCTION INTO
THE SLOPE OF THE TERRAIN.

ABOVE: VIEW FROM ABOVE, OF
THE HILL TOWARDS THE CITY;

BELOW: VIEW OF THE EXISTING
NEIGHBORHOOD TOWARDS THE
CORNICHE AND THE FOREST.

one's own senses. Instead, perceived density — for example from the Rue des Dames Blanches, or from the route on the crest — is what will be important and discussed; the type of urban space proposed, and its proportions; and the quality of open spaces and of the different urban forms that forcefully modify the perception of density.

The three scenarios propose substantially different approaches with respect to perceived density — a matter that is explicit in the elaboration of the three spatial configurations. In the first scenario, continuity with the existing fabric inserts the project in the current urban situation and reduces the potential shock; in the second, the project becomes an opportunity to open large windows on the landscape; in the third, the relationship of osmosis with the topography has an effect on the notion of perceived density.

Introspection.   Among the inhabitants who live around the Rue des Dames Blanches, some belong to the "happy few" living in large villas well protected by walls and high hedges, located mostly to the east of the site. If we look into the character of the new proximity between the social housing project and these large properties without facile irony, but at the same time without dramatic emphasis, we come to the conclusion that it is important to carefully consider this relationship.

Therefore, all the scenarios explore the consequences of the new buildings in terms of introspection towards the existing lots; each scenario defines spaces of mediation through which to make the transition between the different parts. While many sociological research projects point to the presence of "airlocks" between zones with different income levels, our position is that it is preferable to assign this role of mediation of differences to different types of space, instead of a situation of multiple barriers.

Ways of living together and urban forms.   The first scenario proposes living in a fabric composed, for the most part, of individual homes with gardens or patios. As a result most of the space is private, and the slope of the land permits each home to have an unencumbered view to the west and the lower part of the area.

The second scenario proposes living in a collective way, in apartments with terraces, loggias, or small gardens. From the large green areas, views open crosswise towards the forest and city. The forest can be extended all the way to the Rue des Dames Blanches.

The third scenario, finally, posits living in close contact with nature, in a fabric that is inserted and almost concealed in the terrain. Once again, the public spaces can be utilized by the inhabitants, but in different ways with respect to the first scenario, determining a situation that is simultaneously intimate and rich in public spaces.

The average size of the housing units is 80 square meters, and great typological variety, in all the scenarios, permits the exploration of extremes: large homes for big families (120-130 m^2) and small ones (50-60 m^2) for other populations, such as senior citizens, for example.

The reflection on ways of living and urban forms takes the viewpoint of the inhabitants of the new development who are not present in the debate but will be its direct users. Our role is to represent, in the project, the physical dimension of their wellbeing, their needs, desires, and dreams.

Sustainable development. Energy, water, and biodiversity.   Strategies for sustainable development lie behind all three scenarios: from proper orientation of the buildings and the control over shadows to the gathering of rainwater and reinforcement of biodiversity. Only a part of the strategies connected with sustainable development has considerable consequences on construction costs. The scenarios explore the range of possibilities and, in particular, the outcome of different configurations on issues of consumption of energy and water. The subjects implicated, in this case, are not only the inhabitants of the development and its vicinity, or of Brussels, but those of a larger entity – in general, all those who are concerned about a long-term future.

A concept, an icon.   Each scenario has been developed and analyzed in relation to the themes outlined above, in which the viewpoints of the subjects and actors involved are also represented. A concept expresses the main hypothesis, and the reference to a work (icon) helps to indicate its salient characteristics. Each of the themes described in the above paragraphs has contributed to the construction of the scenario, and together with the others determines a grid of interpretation of the proposed configurations. The grid also suggests how to organize the debate.

The project, presented on two occasions to juries put together in different ways, reached a maximum score in both cases. However the final result, based on different criteria, reversed the judgment of the experts.

STREETS, ROUTES, FACILITIES
2,500 LODGINGS/HA;
3,700 LODGINGS/HA.

WATER COLLECTION BASIN AND FOREST
DITCH
PUBLIC GARDENS
PRIVATE GARDENS

0 10   50   100   200m

SCENARIO 01: INHABITED FABRIC.

STREETS, ROUTES, FACILITIES
2,500 LODGINGS/HA;
3,700 LODGINGS/HA.

DITCH AND COLLECTION BASIN
LANDSCAPE COLLECTION BASIN
GREEN SPACE
PUBLIC GARDENS
PRIVATE GARDENS

0 10   50   100   200m

SCENARIO 02: INHABITED STRIPS.

STREETS, ROUTES, FACILITIES
2,500 LODGINGS/HA;
3,700 LODGINGS/HA.

WATER COLLECTION BASIN AND FOREST
DITCH
PUBLIC GARDENS
PRIVATE GARDENS

0 10   50   100   200m

SCENARIO 03: INHABITED LEAF.

Many scholars recognize a new interest in the tool of the scenario which has re-emerged in recent years (HOPKINS, ZAPATA 2007). The phenomenon is vast and points to the need to think about the future using forms of thought, technical and rhetorical constructs, of a different type: from "visioning" the description of a desired future based on a process of exchange and sharing; to forecasting, which investigates probabilities; to the scenario, which explores possibilities. These ways of thinking about the future take on meaning if they are inserted in a discussion in which the construction of the project is part of a "continuing, deliberative work among multiple actors" (HOPKINS, ZAPATA 2007: 6). This is particularly true of visioning, when it is not restricted to a few repetitive clichés. If I have overlooked "vision" in this part of the book, it is intentional. It is because even more than the scenario (though many of the reflections of the French "prospective" move in this direction), it makes sense only if it is constructed collaboratively. It cannot take form within reflection limited to design, and even if this were to occur, the vision can exist only if it is adopted by others. Therefore, the study of visioning belongs to a framework of references that is partially different from the one I have utilized.

What remains today is the strong presence of all the different tools for "thinking the future" and, in particular, of the scenario as a tool of construction of the project and development of new knowledge. The utopia contained in the risk society is that of a "responsible modernity" (BECK 1992) that allows us to pass from hybridization to a new, radicalized, reflexive modernity, concerned with its unintentional and secondary consequences. It is sustained by a different form of rationality that accepts the high level of interpenetration of ecological problems, the fact that they cannot be broken down into separate parts or reduced. It accepts their variability in time and space, their uncertainty, and their status as collective and spontaneous problems (DRYZEK 1987).

Not knowing how to forecast, we try to grasp the possible magnitude of certain phenomena and their consequences by constructing *trend scenarios*; we try to influence the future by organizing consensus around *repulsive scenarios*, or, on the other hand, *auspicable scenarios*; we try to innovate by producing *scenarios of contrast* that do not reproduce trends, but propose new landscapes and spaces.

The construction of scenarios lays the groundwork for narrations and stories, plural accounts we can perhaps re-read in the future as symptoms of an epochal passage. If the descriptive urge and large number of descriptions of the contemporary territory have constructed a discourse coherent with the thesis of the end of the grand narratives (LYOTARD 1979), the urgency of imagining the future as a collective act lays the groundwork for new narrations, which are "grand" precisely because they are collective. In this sense, descriptions and scenarios are discursive formations on the contemporary age in the same manner: they reveal its movements and anxieties, indicating signals and possible trajectories.

All the major authors – along with Berger, Jouvenel, or H.G. Wells – underscore the need for new social institutions, for forums in which scenarios can be debated, for research centers of the future. They all insist on the role of the construction of scenarios in processes of self-organization, adaptation, and innovation: "desire as the productive force of the future is also the main driving force of self-organization", writes Godet in the introduction to the text by Hatem on the *prospective* (HATEM 1993: 12).

A new narrative is under construction today in urbanism. It is the story of dissolving the project of the modern city and replacing it, through trial and error, with a new order, new visions or, more simply, this is a reflection on the design of cities and territories, on its argumentative and heuristic role in the definition of the future of the city.

THE PROJECT AS KNOWLEDGE PRODUCER

Epilogue

Having reached the conclusion of this book, I have the impression of having crossed many territories, but also at times of having proceeded at an exceedingly fast, light pace. There are many places where I would have preferred to linger, so I think of each of them as the point of application of new questions, open to further research or to contact with research already conducted or in progress. I would have liked to delve deeper in to the tradition of design research, the production of concepts in the design of the city, or relational construction and that of situated knowledge. If I have failed to do so, it is only because I felt it was urgent to clarify my viewpoint – not out of arrogance or closed-mindedness – and contribute to a debate that rethinks it, possibly disproves it, or simply uses it as a path leading to other paths.

Design activity is important for the knowledge it produces, and awareness of its role goes hand in hand with awareness of its responsibility. It may be the importance of this hypothesis that has prompted me to see forms of continuity in the reflection on the design of the city through time, with an ongoing focus on the production of ideas and interpretations. Contrasting and contradictory worlds, differences, and ruptures do exist; this book in some ways takes a stance of profound opposition to much of recent design culture. I have moved in time, questioning the history of the design of cities and territories starting from their project – a selective and instrumental itinerary that may appear to be fragmentary and partisan. Maybe it is: it is an attempt to comprehend epistemologies kept at the margins.

In the first part I have taken some risks: thinking about the production of concepts inevitably links back to a history of ideas that is separated from subjects, atemporal. The "autonomy" of the artwork, the *concept* investigated in the three chapters, evokes a structuralism separated from experience and practices. Nevertheless, concepts are also a form of concrete action in space; they interpret and abstract, at times they generalize; they are tools, like the blade of a knife, that make it possible to cross layers of differences; and they offer points of escape.

In the second part I could – and should – have granted more space to the rise, in relation to processes of individualization, of the theme of difference on which a large part of design has been constructed, starting in the 1960s. Along this line I opted to observe the surfacing in a part of geographical thought from the 1800s to the 1900s. Then Deleuze and Guattari [1980] return on the fundamental participation of the *milieu* in the construction of the form, and they meet with many exegetes and banal interpretations. I am interested in the capacity of the project to interpret and describe, making itself into a way of taking a stand through space, and not merely serving as a refuge or an expedient.

In the third part, which discusses the project for future thinking, I could have left more room for the conceptual instruments required for establishing relations between different times, the paradoxical construction (again), the role of error. I could have more explicitly evoked Popper on the conjectural nature of any theory, to sustain the relevance of a project that takes the incursion of the future into urban thinking as the basis of production of an original form of knowledge. I believe, however, that design activity produces concepts, and therefore its relationship to philosophy is integral, though it is full of intersections.

The theme is starting over. This takes place not from a *tabula rasa*, as Deleuze writes in the dialogues with Claire Parnet (1977), from the *"recherche d'une première certitude"*, but "taking up the interrupted line, to join a segment to the broken line, to make it pass between two rocks in a narrow gorge, or over the top of a void where it had stopped".

AA.VV. 1968, *Idee per la città comunista*, il Saggiatore, Milano (orig.: 1966, in Russian, *Novyi element rasselenia. Na puti k novomu gorodu*, English translation of Italian edition: 1971, *The Ideal Communist City*, i Press Series on the Human Environment, trans. Renee Neu Watkins, New York: George Braziller)

AA.VV. 'Il concorso per il quartiere residenziale alle Barene di S. Giuliano, Venezia-Mestre', *Casabella*, n. 242, 1960

AA.VV. 2000, *Mutations*, Barcelona: Actar

AA.VV. 1963, *Traffic in Towns, A study of long term problems of traffic in urban areas*. Reports of the Steering Group and Working Group appointed by the Minister of Transport, London: Her Majesty's Stationery Office

AA.VV., 2003, *USE (Uncertain States of Europe)*, Milano: Skira

Abercrombie P., Forshaw J.H. 1943, *County of London Plan*, London: MacMillan

Abercrombie, P. 1944, *Greater London Plan*, London: His Majesty's Stationery Office, 1944

Adam, J.M. 1993, *La Description*, *Que sais-je*, Paris: Presses Universitaires de France

Alexander, C. 1965, 'The City is not a Tree', in *Architectural Forum*, vol. 122, n. 1, pp 58-62 (Part I), vol. 122, n. 2, pp. 58–62 (Part II), now also in http://www.patternlanguage.com/archives/alexander1.htm

Alexander, C. 1971, *Notes on the Synthesis of Form*, Cambridge: Harvard University Press

Aligica, P.D. 2007, *Prophecies of Doom and Scenarios of Progress*, New York, London: Continuum

Amin A., Thrift N. 2001, *Cities. Re-imagining the Urban*, Cambridge: Polity Press

Amphoux P., Grosjean G., Salomon J. 2001, *La densité urbaine: du programme au projet urbain*. Rapport de recherche n. 142, Lausanne: IREC Institut de Recherche sur l'Environnement Construit

Andler, D. 1987, *Problème - Une clé universelle?*, in Stengers I., ed., *D'une science à l'autre. Des concepts nomades*, Paris: Seuil

Ardigò, A. 1967, *La diffusione urbana*, Roma: Ave-an Veritas Ed.

Arminius (Adelheid Gräfin Dohna-Poninski) 1874, *Die Großstädte in ihrer Wohnungsnot und die Grundlagen einer durchgreifenden Abhilfe*, Leipzig

Arnheim, R. 1969, *Visual Thinking*, Berkeley-Los Angeles: Regents of University of California

Ascher, F. 2001, *Les nouveaux principes de l'urbanisme*, Paris: Éditions de l'Aube

Augé, M. 1986, *Un ethnologue dans le métro*, Paris: Hachette

Aureli, P.V. 2004, *La città arcipelago e il suo progetto*, doctoral thesis at IUAV University of Venice

Avermaete, T. 2005, *Another Modern*, Rotterdam: Nai Publishers

Bacon, E. N. 1967, *Design of Cities*, Viking Penguin Inc. (revised edition: 1974, Penguin Books 1976)

Bagnasco, A. 1977, *Tre Italie. La problematica territoriale dello sviluppo urbano*, Bologna: Il Mulino

Balducci, A. 1991, *Disegnare il futuro*, Bologna: Il Mulino

Baljon, L. 1992, *Designing Parks*, Amsterdam: Architectura et Natura Press

Banham R., Barker P., Hall P., Price C. 1969, 'Non-plan: an experiment in freedom', *New Society*, 20 March, London: New Society Publ.

Barattucci, C. 2004, *Urbanizzazioni disperse. Interpretazioni e azioni in Francia e in Italia, 1950–2000*, Roma: Officina Edizioni

Barthes, R. 1972, *The Structuralist Activity*, in *Critical Essays*, Evanston, IL: Northwestern University Press (first publication in French, 1963: *Lettres Nouvelles*)

Barthes, R. 2002, *Comment vivre ensemble. Notes de cours et de séminaires au Collège de France, 1976–1977*, Paris: Seuil Imec (Eng. 2013, *How to Live Together: Novelistic Simulations of Some Everyday Spaces*, New York: Columbia University Press)

BAUER G., Roux J.M. 1976, *La rurbanisation, ou la ville éparpillée*, Paris: Seuil

BECATTINI, G. 1997, *Prato nel mondo che cambia (1954–1993)* in *Prato. Storia di una città*, vol. 4 *Il distretto industriale (1943–1993)*, Firenze: Le Monnier

BECATTINI, G. 1998, *Distretti industriali e made in Italy,* Torino: Bollati Boringhieri

BECK, U. 1986, *Risikogesellschaft - Auf dem Weg in eine andere Moderne*, Frankfurt a/M: Suhrkamp (Eng. trans.: 1992, *Risk Society: Towards a New Modernity*, London: Sage Publications)

BECK U., GIDDENS A., LASH S. 1994, *Reflexive Modernization*, Cambridge: Polity Press

BELL, D. 1967, 'Notes on the Post-Industrial Society', in *Public Interest*, winter and spring

BELLAMY, E. 1888, *Looking Backward, 2000–1887*, Boston: Houghton, Mifflin and Co.

BENJAMIN, W. 1936, *Das Kunstwerk im Zeitalter seiner technischen Reproduzierbarkeit* (Eng. trans.: *The Work of Art in the Age of Mechanical Reproduction*, 2008, trans. A.J. Underwood, London: Penguin Books)

BENJAMIN, W. 1962, *Angelus Novus - saggi e frammenti*, Torino: Einaudi Berdoulay, V. 1993, *La géographie vidalienne: entre texte et contexte*, in Claval P., ed., *Autour de Vidal de la Blache*, Mémoires et documents de Géographie, Paris: CNRS Éditions

BERGER, G. 1964, *Phénoménologie du temps et prospective*, Paris: Presses Universitaires de France

BIANCHETTI, C. 1987, *Teorie dell'indagine e processi cognitivi nella pianificazione territoriale*, in *Archivio di Studi Urbani e Regionali*, a. XVIII – n.29

BIANCHETTI, C. 1989, *Conoscenze e piano*, Milano: Franco Angeli

BIANCHETTI, C., ed. 2004, *Territori sempre più simili*, Piano Progetto Città n.22–23

BOERI S., LANZANI A., MARINI E. 1993, *Il territorio che cambia,* Milano: Editrice Abitare Segesta

BOLTANSKI L., THÉVENOT L. 1991, *De la justification, Les économies de la grandeur*, Paris: Gallimard

BORRET, K. 1994, 'Organizing urban space in a peripheral environment. The project for the bay-area at Rouen by the Projectteam Stadsontwerp', in *Quaderno della ricerca sulle trasformazioni dell'habitat contemporaneo* n. 2, Venezia: IUAV

BOTAR, O. 1998, *Prolegomena to the Study of Biomorphic Modernism: Biocentrism, László Moholy-Nagy's New Vision and Erno Kallai's Bioromantik*, PhD thesis, Graduate Department of the History of Art, University of Toronto

BOYER, M.C. 2003, *Cognitive Landscapes*, in Spellman C., ed., *Re-Envisioning Landscape/Architecture*, Barcelona: Actar

BOUDON, Ph. 1975, *Architecture et architecturologie*, Paris: A.R.E.A

BOUDON, Ph. 2004, *Conception*, Paris: Éditions de la Villette

BOULEZ, P. 2005, *Leçons de musique, Points de repère*, vol. 3, Paris: Christian Bourgois

BOZZUTO P., COSTA A., FABIAN L., PELLEGRINI P. 2008, *Storie del futuro, 04*, Roma: Officina

BRANZI, A. 2007, *Modernità debole e diffusa*, Milano: Skira

BROADBENT, G. 1973, *Design in Architecture* (from the edition of 1988, London: David Fulton Publishers)

BROADBENT, G. 1990, *Emerging Concepts in Urban Space Design*, London, New York: Van Nostrand Reinhold International

BRUNET, R. 2002, 'Lignes de force de l'espace européen', *Mappemonde* n. 66

BURNHAM, D.H. 1905, *Report of D.H. Burnham on the Improvement and Adornment of San Francisco*, ed. O'Day E.F., San Francisco: Sunset Press

CACCIARI, M. 1997, *L'arcipelago*, Milano: Adelphi

CANDILIS G., JOSIC A., WOODS S. 1965, *Recent Thoughts in Town Planning and Urban Design*, in Smith Morris E., *New Urban Design Concepts, Greenways and Movement Structures in the Philadelphia Plan*, in Lewis D., ed., *The Pedestrian in the City*, London: Elek Books

CAVALLETTI, A. *La città biopolitica: mitologie della sicurezza*, Milano: Bruno Mondadori

CAZES, B. 1986, *L'histoire des futurs*, Paris: Éditions Seghers

CHAPUIS, J.Y. 2003, 'De la ville historique à la ville archipel', *Etudes Foncières* n. 105

CHOMBART DE LAUWE P.H., ANTOINE S., BERTIN J., COUVREUR L., GAUTHIER J. 1952, *Paris et l'agglomération parisienne*, Paris: PUF

CHRISTENSEN, K.S. 1985, 'Coping with Uncertainty in Planning', *Journal of the American Planning Association,* vol. 51, n. 1

CLAVAL P., CLAVAL F. 1981, *La logique des villes*, Paris: Litec

CLEMENTI A., DE MATTEIS G., PALERMO, P.C. 1996, ed., *Le forme del territorio italiano*, Bari: Laterza

COLLINS, G.R. 1968, *Lo sviluppo della pianificazione lineare*, introduction to: Soria y Mata, A. *La città lineare*, Milano: il Saggiatore

COLQUHOUN, A. 1972, 'Displacement of Concepts in Le Corbusier', in 1981, Chap. 2, Essays in Architectural Criticism, Cambridge: MIT Press

COMMISSION ON THE YEAR 2000. 1967, *Toward the Year 2000: work in progress*, part of *Daedalus: proceedings of the American Academy of Arts and Sciences*, Boston, MA

CONFURIUS, G. 2000, editorial, *Daidalos, Diagrammania* n. 74

COOKE, C. 2001, *Extensive or Intensive Development? A Century of Debates and Experience in Moscow*, in Echenique M., Saint A., ed., *Cities for the New Millennium*, London: Spon Press

CORBOZ, A. 1998, *Ordine sparso*, Milano: Franco Angeli

CORNER J., MACLEAN A.S. 1996, *Taking Measures Across the American Landscape*, New Haven and London: Yale University Press

CORNER, J. 1999, 'The Agency of Mapping: Speculation, Critique and Invention', in Cosgrove D. *Mappings*, London: Reaktion Books

COSGROVE, D. 1999, *Mappings*, London: Reaktion Books

COYNE, R. 2005, 'Wicked problems revisited', *Design Studies* 26:1

CROSS, N. 2006, *Designerly Ways of Knowing*, London: Springer

CRYSLER, G. 2003, *Writing Spaces: Discourses of Architecture, Urbanism and the Built Environment, 1960–2000*, New York and London: Routledge

DAHAN DALMEDICO, A. 2007, ed., *Les modèles du futur*, Paris: La Découverte

DAVIDSON, D. 1984, 'On the Very Idea of a Conceptual Scheme', in *Inquiries into Truth and Interpretation* (from the 2001 edition, Oxford: Oxford University Press)

DE CARLO, G. 1964, *Questioni di architettura e urbanistica*, Urbino: Argalia

DE CAUTER, L. 2005, *Capsular Civilisation: The City in an Age of Fear*, Rotterdam: NAi

DEHAENE, M. 2002, *A Descriptive Tradition in Urbanism – Patrick Abercrombie and the Legacy of Geddesian Survey*, PhD thesis, KU Leuven

DELEUZE, G. 1977, *Dialogues* (with C. Parnet), Paris: Flammarion

DELEUZE, G. 1986, *Foucault*, Paris: Éditions de Minuit

DELEUZE G., Guattari F. 1980, *Mille Plateaux*, Paris: Éditions de Minuit

DELEUZE G., Guattari F. 1991, *Qu'est-ce que la philosophie?*, Paris: Éditions de Minuit

DEWEY, J. 1920, *Reconstruction in Philosophy*, expanded edition, 1948, Boston: Beacon Press

DEWEY, J. 1939, *Logic: The Theory of Inquiry*, New York: Holt, Rinehart and Winston

DICKINSON, R.E. 1964, *City and Region*, London: Routledge & Kegan Paul Ltd

DIENER, R. et al. 2006, *Die Schweiz. Ein städtebauliches Portrait*, Basel: Birkhäuser

DOYLE, A. C. 1902, *The Hound of the Baskervilles*, from the edition of 1981, London: Penguin Books

DRYZEK, J.S. 1987, *Rational Ecology: Environment and Political Economy*. New York: Basil Blackwell

EBERSTADT, R. 1909, *Handbuch des Wohnungswesens und der Wohnungsfrage*, Jena: G. Fischer

ECO, U., SEBEOK, T.A. 1983, *Il segno dei tre, Holmes, Dupin, Peirce*, Milano: Bompiani

ECO, U. 1990, *I limiti dell'interpretazione*, Milano: Bompiani

EISENMAN, P. 1998, 'Diagram: An Original Scene of Writing', *ANY* n. 23, *Diagram Work*

EISINGER A., SCHNEIDER M., ed., s.d. *Urbanscape Switzerland*, Avenir Suisse, Basel, Boston, Berlin: Birkhäuser

ELLIN, N. 1996, *Postmodern Urbanism*, Cambridge, MA; Oxford: Blackwell

ENGELS, *Antidühring*, 1878

FABOS J., MILDE G.T., WEINMAYR V.M. 1968, *Frederick Law Olmsted Sr: Founder of Landscape Architecture in America*, Cambridge: The University of Massachusetts Press

FEIN, A. 1972, *Frederick Law Olmsted and the American Environmental Tradition*, New York: George Braziller

FERRARO, G. 1998, *Rieducazione alla speranza: Patrick Geddes planner in India, 1914–1924*, Milano: Jaca Book

FISCHER, R. 1935, *The Design of Experiments*, Edinburgh: Oliver and Boyd

FLECHTHEIM, O.K. 1972, 'Futurology – The New Science of Probability?', in Toffler A., *The Futurists*, Random House

FOGEL, R. 1964, *Railroads and American Economic Growth: Essays in Econometric History*, Baltimore: Johns Hopkins University Press

FORD, H. (in collaboration with Crowther, S.) 1922, *My Life And Work*, New York: Garden City Publishing Company, Inc.

FORESTIER, J.C.N. 1908, *Grandes villes et systèmes de parcs* (reissued 1997, Paris: Norma)

FOUCAULT, M. 1971, *L'ordre du discours*, Paris: Gallimard

FOUCAULT, M. 1975, *Surveiller et punir*, Paris: Gallimard

FOUCHIER, V. 1994, 'Penser la densité', *Etudes Foncières* n. 64

FOUCHIER, V. 1998, *Les densités urbaines et le développement durable: le cas de l'Île de France et des villes nouvelles*, Paris: Secrétariat général du groupe central des villes nouvelles

FORMAN R. T. T., GODRON M. 1986, *Landscape Ecology*, New York: John Wiley

FORMAN, R. T. T. 1995, *Land Mosaics: the Ecology of Landscapes and Regions*, Cambridge University Press, 1995

FUÀ G., ZACCHIA C. 1983, ed., *Industrializzazione senza fratture*, Bologna: Il Mulino

GADAMER, H.G. 1960, *Wahrheit und methode*, Tübingen: J. C. B. Mohr Verlag

GALBRAITH, J.K. 1958, *The Affluent Society*, Boston: Houghton Mifflin Company

GEDDES, P. 1915, *Cities in Evolution*, London: Williams & Norgate

GEERZT, C. 1973, *The Interpretation of Cultures*, New York: Basic

GERMANI, G. 1991, *Saggi sociologici*, Napoli: Edizioni Libreria dell'Ateneo

GIL, F. 1978, *Enciclopedia Einaudi*, entry *Conoscenza*, Torino: Einaudi

GINZBURG, C. 1992, *Miti emblemi spie. Morfologia e storia*, Torino: Einaudi

GIRARD, C. 1986, *Architecture et concepts nomades*, Bruxelles: Pierre Mardaga

GLOEDEN, E. 1923, *Inflation der Grosstädte*, Berlin: Der Zirkel, Architecktur Verlag

GODET, M. 1992, *De l'anticipation à l'action*, Dunod

GOODMAN, N. 1968, *Languages of Art: An Approach to a Theory of Symbols*. Indianapolis: Bobbs-Merrill

GOULD, S.J. 1980, *The Panda's Thumb*, New York: W. W. Norton

GRAHAM S., Marvin S. 2001, *Splintering Urbanism*, New York: Routledge

GRACQ, J. 1985, *La forme d'une ville,* Paris: Corti

GREGOTTI, V. 1966, *La forma del territorio*, Edilizia Moderna, n. 87–88

GUTKIND, E.A. 1953, *The Expanding Environment: The End of Cities – The Rise of Communities*, London: Freedom Press

GUTKIND, E.A. 1962, *The Twilight of Cities*, New York: Free Press of Glencoe, MacMillan Company

HABRAKEN, N. J. 1962, *De Dragers en de Mensen. Het einde van de massa Woningbouw*, Amsterdam: Scheltema en Holkema (Eng. trans.: 1972, *Supports: an Alternative to Mass Housing*, London: The Architectural Press and New York: Praeger

HABRAKEN, N. J. 1998, *The Structure of the Ordinary*, London, Cambridge MA: MIT Press

HARDINGHAM, S. 2003, ed., *Cedric Price: Opera*, Chichester: Wiley-Academy

HATEM, F. et al. 1993, *La prospective. Histoire, pratiques et méthodes*, Paris: Economica

HEGEMANN, W. 1930, *Das Steinerne Berlin*, Berlin: Kiepenhauer

HEYNEN H., VANDERBURGH, D. 2003, *Inside Density*, Bruxelles: La Lettre Volée

HILBERSEIMER, L. 1944, *The New City – Principles of Planning*, Chicago: Paul Theobald

HILBERSEIMER, L. 1949, *The New Regional Pattern: Industries and Gardens, Workshops and Farms*, Chicago: Paul Theobald

HOPKINS, L.D., ZAPATA, M., ed. 2007, *Engaging the Future, Forecasts, Scenarios, Plans, and Projects*, Cambridge, Massachusetts: Lincoln Institute of Land Policy

HOWARD, E. 1902, *Garden Cities of To-Morrow*, London: S. Sonnenschein & Co., Ltd.

HUET, B. 2003, *Sur un état de la théorie de l'architecture au XX siècle*, Paris: Éditions Quintette

ILSES. 1962, *Nuova dimensione della città: la città regione*, Atti del seminario di Stresa, Milano

INDOVINA, F. 1990, ed., *La città diffusa*, Venezia: Dipartimento di Analisi Economica e Sociale del Territorio, Istituto Universitario di Architettura

INDOVINA F., Fregolent L., SAVINO M., ed., 2005, *L'esplosione della città*, Bologna: Editrice Compositori

INFUSSI, F. 1995, *La costruzione del progetto-norma: tecniche di 'traduzione normativa' e forme di piano*, Dipartimento di Scienze del Territorio – Facoltà di Architettura del Politecnico di Milano

JACKSON, J. B. 1970, *Landscapes: Selected Writings of J. B. Jackson,* Zube E.H., ed., Amherst, MA: University of Massachusetts Press

JAEGGI, A. 1987, *Das Grosssiedlungsmodel: Von der Rationaliserung zum Städtebau* in *Vier Berliner Siedlungen der Weimarer Republik*, exhibition catalogue, Bauhaus-Archiv, Berlin: Argon Verlag

JANSSENS, P., DE WAEL, P. 2005, *50 années de marché immobilier en Belgique 1953–2003*, Bruxelles: Roularta Books

JENCKS, C. 1995, *The Architecture of the Jumping Universe: A Polemic: How Complexity Science is Changing Architecture and Culture*, London: Academy editions, from the revised edition, 1997

JONES, J.C., THORNLEY, D.G. 1963, ed., *Conference on Design Methods*, Oxford: Pergamon Press

JOUVENEL, B. 1964, *L'art de la conjecture*, Futuribles, Monaco: Éditions du Rocher

KAHN, H. 1962, *Thinking About the Unthinkable*, New York: Horizon Press

KAHN, H., WIENER, A.J. 1967, *The Year 2000*, New York: The Macmillan Company

KAHN, H., BRUCE-BRIGGS, B. 1972, *Things to Come – Thinking about the Seventies and Eighties*, New York: The MacMillan Company

KELLER, S.B. 2005, *Systems Aesthetics: Architectural Theory at the University of Cambridge, 1960–1975*. Doctoral thesis, Harvard University

KELLER, S.B. 2007, 'Systems Aesthetics, or How Cambridge Solved Architecture', in T. Anstey, K.; Grillner, R.; Hughes, ed., *Architecture and Authorship*, London: Black Dog Publishing

KEPES, G. 1954, *The New Landscape in Art and Science*, Chicago: Paul Theobald

KHAN-MAGOMEDOV, S. O. 1987, *Pioneers of Soviet Architecture*, Thames and Hudson

KOOLHAAS, R. 1978, *Delirious New York: a Retroactive Manifesto for Manhattan*, New York: Oxford University Press

KOOLHAAS, R. 1993a, 'New Urban Frontiers', in Devolder A., editor-in-chief, *Alexander Polder, New Urban Frontiers*, Bussum: Thoth

KOOLHAAS, R. 1993b, 'Urban Operations', *D: Columbia Documents of Architecture and Theory*, vol. 3

KOOLHAAS R., Mau, B., OMA. 1995, *S,M,L,XL*, Rotterdam: 010 Publishers

KOPP, A. 1967, *Ville et Révolution*, Paris: Anthropos [Eng. trans.: 1970, *Town and Revolution – Soviet Architecture and City Planning 1917–1935*, New York: George Braziller]

KROPOTKIN, P. 1899, *Fields, Factories and Workshops*, Eng. edit. 1974, New York: Harper & Row

KRSTIC, V. 1997, 'Liquid Constructions, Tokyo and the End of Space', in Bognar Botond. *Tokyo*, World Cities Series, John Wiley & Sons

KUHN, T. S. 1962, *The Structure of Scientific Revolutions*, I ed. Chicago: Univ. of Chicago Press

LANZANI, A. 1991, *Il territorio al plurale. Interpretazioni geografiche e temi di progettazione territoriale in alcuni contesti locali*, Milano: Franco Angeli

LAUGIER, M.A. 1753, *Essai sur l'architecture*, Paris: Duchesne

LE CORBUSIER. 1933, *La Ville Radieuse: Eléments d'une Doctrine d'urbanisme pour l'équipement de la Civilisation Machiniste*, Boulogne: Éditions de L'architecture d'aujourd'hui (Collection de l'équipement de la civilisation machiniste)

LE CORBUSIER. 1943, *La Charte d'Athènes*, Paris: Éditions de Minuit

LE CORBUSIER. 1946, *Les trois établissements humains*, Paris: Denoël (Collection Ascoral, vol. 7)

LEFEBVRE, H. 1966, 'Introduction à l'étude de l'habitat pavillonnaire', in Raymond H., Haumont N., Raymond M.G, Haumont A., *L'Habitat pavillonnaire*, Paris: Éditions du C.R.U.

LEFEBVRE, H. 1968, *Le droit à la ville*, Paris: Éditions Anthropos

LÉVI-STRAUSS, C. 1993, *Regarder Ecouter Lire*, Paris: Plon [Eng. trans. 1998, *Look, Listen, Read*, New York: Basic Books]

LIPPITT, R., WATSON, J., WESTLEY, B. 1958, *The Dynamics of Planned Change*, New York: Harcourt Brace & Company

LOBSINGER, M.L. 2000, 'Cedric Price: An Architecture of the Performance', *Daidalos* n. 74

LOI, D. 2000, 'Caractères généraux de la causalité vidalienne et objets de l'explication dans le Tableau de la géographie de la France', in Robic M.C., ed., *Le Tableau de la géographie de la France de Paul Vidal de la Blache*, Paris: CTHS

LOMBARDO, T. 2008, *La città normale*, doctoral thesis, IUAV University of Venice

LONERO, G. 2005, 'Chandigarh prima di Chandigarh: il contributo di Albert Mayer e della sua squadra', *Annali di Architettura* n. 17

LYOTARD, J. 1979, *La Condition postmoderne: Rapport sur le savoir*, Paris: Minuit

MALFROY, S. 2000, *Penser et représenter la ville* (lecture notes of the course in Théorie urbaine, II edit.), EPFL

MANCUSO, F. 1978, *Le vicende dello zoning*, Milano: il Saggiatore

MANGIN, D. 2004, *La ville franchisée: Formes et structures de la ville contemporaine*, Paris: Éditions de la Villette

MANTIA, G. 2005, *Lo spazio poroso*, doctoral thesis in Urban Planning, IUAV University of Venice

MANTZIARAS, P. 2008, *La ville-paysage – Rudolf Schwarz et la dissolution des villes*, Genève: Metis Presses

MARBACH, G., MAZZIOTTA, C., RIZZI, A. 1991, *Le previsioni – Fondamenti logici e basi statistiche*, Milano: EtasLibri

MARQUARD, O. 1987, *Apologie der Zufälligen*, Stuttgart: Reclam

MASSÉ, P. 1965, *Le plan ou l'anti-hasard*, Paris: Gallimard

MASPERO, F. 1990, *Les passagers du Roissy-Express*, Paris: Éditions du Seuil

MATHEWS, S. 2007, *From Agit-Prop to Free Space: The Architecture of Cedric Price*, London: Black Dog Publishing

MEADOWS, D.H., MEADOWS, D.L., RANDERS, J., BEHRENS, W.W. 1972, *The Limits to Growth*, New York: Universe Books

MELLER, H. 1990, *Patrick Geddes – Social Evolutionist and Planner*, London and New York: Routledge

MENDINI, A. 1985, introduction to Venturi R., Scott Brown D., Izenour S., *Imparando da Las Vegas*, Venezia: Cluva editrice

MENNA, F. 1975, *La linea analitica dell'arte moderna. Le figure e le icone*, Torino: Einaudi

MILJUTIN, N. 1930, *Socgorod. Problema stroitel'stva socialisticeskich gorodov* (Eng. trans, 2009, *Sozgorod*, Berlin: Dom Publishers)

MORIER, F. 1994, ed., *Belleville, Belleville, visage d'une planète*, Paris: Creaphis

MORIN, E. 1999, *La tête bien faite*, Paris: Seuil

MUNARIN S., TOSI, M.C. 2001, *Tracce di città*, Milano: Franco Angeli

MURATORI, S. 1960, *Studi per una operante storia urbana di Venezia*, Roma: Istituto Poligrafico dello Stato (first published in *Palladio* n. 3–4, 1959)

MVRDV. 1997, 'Light Urbanism', *Archis* n. 2

MVRDV, Maas, W. 1998, *Farmax: Excursions on Density*, Rotterdam: 010 Publishers

MYERS D., KITSUSE, A. 2000, 'Constructing the Future in Planning: A Survey of Theories and Tools', *Journal of Planning Education and Research* 19(3)

NEUTELINGS, W.J. 1990, 'Fragmentatie in de periferie: de 'tapijtmetropool', *Archis* n. 3

NOLEN, J. 1916, *City Planning*, New York and London: D. Appleton and Company

OCKMAN, J. 2000, ed., *The pragmatist Imagination*. Princeton: Princeton Architectural Press

OLMSTED, F.L. 1870, *Public Parks and the Enlargement of Towns*, reprint 1970, New York: Arno Press and the *New York Times*

OLMSTED, F.L. 1971, *Civilizing American Cities: A Selection of Frederick Law Olmsted's Writings on City Landscapes* (ed. S. B. Sutton), Cambridge, MA: MIT Press

OTAM (Omnium Technique d'Aménagement) 1971a, *Une image de la France en l'an 2000, Travaux et Recherches de prospective*, n. 20, Paris: La Documentation française

OTAM (Omnium Technique d'Aménagement) 1971b, *Scénarios d'Aménagement du Territoire, Travaux et Recherches de prospective* n. 12 Paris: La Documentation Française

OTTANI CAVINA, A. 1994, *I paesaggi della ragione*, Torino: Einaudi

PACI, M. 2005, *Nuovi lavori, nuovo welfare*, Bologna: Il Mulino

PANIZZA M., CASTALDINI, D., PELLEGRINI, M., GIUSTI, C., PIACENTINI, D. 2004 *Matrici geo-ambientali e sviluppi insediativi, una ipotesi di ricerca*, in *Per un Atlante Storico Ambientale Urbano*, Mazzeri, C., ed., Modena: Edizioni APM, series *Le città sostenibili*

PANZINI, F. 1993, *Per i piaceri del popolo. L'evoluzione del giardino pubblico in Europa dalle origini al XX secolo*, Bologna: Zanichelli

PARK, R.E., BURGESS, E.W., MCKENZIE, R.D. 1925, *The City*, (edit. 1967, Chicago: University of Chicago Press)

PARKINS, M.F. 1953, *City Planning in Soviet Russia*, Chicago: University of Chicago Press

PASOLINI, P. 1979, *Descrizioni di descrizioni*, ed. Chiarcossi G., Torino: Einaudi

PEACOCKE, C. 1992, *A Study of Concepts*, London, Cambridge, MA: MIT Press

PELLEGRINI, P., VIGANÒ, P. 2006 ed., *Comment vivre ensemble, Q3*, Roma: Officina

PELLINI, U. 1996, Alberi nella storia di Reggio, Reggio Emilia: AGE grafico editoriale

PETITIER, P. 2000, 'D'un Tableau l'autre. Le Tableau de la France de Michelet et le Tableau de la géographie de la France de Vidal de la Blache', in Robic M.C., ed., *Le tableau de la géographie de la France de Paul Vidal de la Blache*, Paris: CTHS

POPPER, K. 1963, *Conjectures and Refutations*, London: Routledge and Keagan Paul

PICCINATO, G., QUILICI, V., TAFURI, M. 1962, 'La città territorio, verso una nuova dimensione', *Casabella Continuità* n. 270

PICCINATO, G. 1974, *La costruzione dell'urbanistica. Germania 1871–1914*, Roma: Officina

PICCINATO, G., DE LUCA, G. 1983, 'Verso una nuova città? Analisi dei processi di diffusione urbana', *Oltre il ponte* n. 2

PICCINATO G., SARTORE, M. 1990, *Spazio rurale e urbanizzazione diffusa: il caso veneto*, in Clementi A., Perego F., ed., *Eupolis- La riqualificazione delle città in Europa*, vol. I, *Periferie oggi*, Roma-Bari: Laterza

POLAK, F. L. 1973, *The Image of the Future*, Amsterdam, London, New York: Elsevier

PRICE, C. 1977, 'More operational originality, less three-dimensional ingenuity, more social structures', *The Architects' Journal* 21–28

PRICE, C. 2003, *The Square Book*, Chichester: Wiley-Academy

PROVINCIA AUTONOMA DI TRENTO, 1968, *Piano urbanistico del Trentino*, Padova: Marsilio

RATTENBURY, K., HARDINGHAM, S. 2007, *SuperCrit #2: Robert Venturi and Denise Scott Brown, Learning from Las Vegas*, London: Routledge

RECLUS, É. 1866, 'Du sentiment de la nature dans les sociétés modernes', *La Revue des Deux Mondes*, vol. 63

RECLUS, É. 1895, 'La Cité du bon accord', *The Evergreen: A Northern Seasonal. The Book of Autumn*, 'published in the Lawnmarket of Edinburgh by Patrick Geddes and colleagues...'

REPS, J.W. 1965, *The Making of Urban America*, New Jersey: Princeton University Press

RITTEL, H.W.J., Webber, M.M. 1973, 'Dilemmas in a General Theory of Planning', *Policy Sciences* n. 4

ROBERT-DEMONTROND, P. 2005 ed., *L'importation de concepts*, Rennes: Éditions Apogée

ROBIC, M.C. 1993, 'L'invention de la "Géographie humaine" au tournant des années 1900: les Vidaliens et l'écologie', in Claval P., *Autour de Vidal de la Blache*, Paris: CNRS Éditions

ROBIC, M.C. 1998, 'Ville et région dans les échanges transatlantiques entre géographes de la première moitié du XX siècle: convergences et diversité des expériences', *Finisterra* 33, n. 65

ROBIC, M.C. 2000, 'Territorialiser la nation. Le Tableau entre géographie historique, géographie politique, géographie humaine', in Robic, M.C., ed., *Le tableau de la géographie de la France de Paul Vidal de la Blache. Dans le labyrinthe des formes*, Paris: CTHS

ROSS, K. 1996, *Fast Cars, Clean bodies. Decolonization and the Reordering of French Culture*, London, Cambridge, MA: MIT Press

ROSSI, P. 1987, ed., *Modelli di città*, Torino: Einaudi

ROWE, P. G. 1987, *Design Thinking*, London, Cambridge, MA: MIT Press

RUSSELL, B. 1948, *Human Knowledge: Its Scope and Limits,* New York: Simon and Schuster

SAARINEN, E. 1943, *The City: Its Growth. Its Decay. Its Future*, New York: Reinhold Publishing

SARTORE M. 1988, 'Forme e processi di Urbanizzazione diffusa. Un'analisi della morfologia insediativa in un'area rurale del Veneto centrale', *Archivio di Studi Urbani e regionali* n. 32

SARTRE, J.P. 1936, *L'imagination* (from the edition of 1956, Paris: Presses Universitaires de France)

SCARPA, L. 1983, *Martin Wagner e Berlino*, Roma: Officina Edizioni

SCHÖN, D.A. 1983, *The Reflective Practitioner*, New York: Basic Books

SECCHI, B. 1984, *Il racconto urbanistico*, Torino: Einaudi

SECCHI, B. 1986, 'Progetto di suolo', *Casabella* n. 521

SECCHI, B. 1988, 'La macchina non banale: una postfazione', *Urbanistica* n. 92

SECCHI, B. 1989, *Un progetto per l'urbanistica*, Torino: Einaudi

SECCHI, B. 1991, 'La periferia', *Casabella* n. 583

SECCHI, B. 1992a, 'Urbanistica descrittiva', *Casabella* n. 588

SECCHI, B. 1992b, 'Un principio di responsabilità', *Casabella* n. 589

SECCHI, B. 1994, *Tre piani, La Spezia, Ascoli, Bergamo*, ed. Bianchetti C., Milano: Franco Angeli
SECCHI, B. 2000, *Prima lezione di urbanistica*, Roma-Bari: Laterza
SECCHI, B. 2005, *La città del XX secolo*, Roma-Bari: Laterza
SECCHI, B. 2008, *La città normale*, in Mittner, D., ed., *La città reticolare e il progetto moderno*, Milano: Città Studi
SECCHI, B., BIANCHETTI, C., VIGANÒ, P. 1998, *Piano Territoriale della Provincia di Pescara*, Provincia di Pescara
SECCHI B., LANZANI, A., VIGANÒ, P. 1993, *Piano Territoriale di Coordinamento di La Spezia e Val di Magra*, Provincia di La Spezia
SECCHI, B., VIGANÒ, P. 1998, 'Un programma per l'urbanistica', in *Urbanistica* 111, July–December
SECCHI, B., VIGANÒ, P. 2009, *Antwerp, Territory of a New Modernity*, Amsterdam: SUN
SECCHI, B., VIGANÒ, P., with FABIAN, L., COSTA, A. 2004, *Scenari retroattivi per il territorio di Modena: la storia si fa con i 'se'*, in *Per un Atlante Storico Ambientale Urbano*, Mazzeri, C., ed. Modena: Edizioni APM, series *Le città sostenibili*
SHANNON, K. 2004, *Rhetorics and Realities, Addressing Landscape Urbanism, Three Cities in Vietnam*, doctoral thesis, Katholieke Universiteit, Leuven
SICA, P. 1978, *Storia dell'urbanistica. Il Novecento*, Roma-Bari: Laterza
SIERKS, H. L. 1926, *Wirtschaftlicher Städtebau*, Dresden: Kaden & Comp.
SIERKS, H. L. 1929, *Grundriss des sicheren, reichen, ruhigen stadt*, Dresden: Kaden & Comp.
SIEVERTS, T. 1997, *Zwischenstadt* (Eng. trans.: 2003, *Cities without Cities: An Interpretation of the Zwischenstadt*, London: Spon Press, New York: Routledge)
SIMON, H. 1969, *The Science of the Artificial*, Cambridge, MA: MIT Press
SMETS, M. 1993, *Aménagement des Quartiers Ouest*, final report of the study conducted by Projectteam Stadsontwerp, Ville de Rouen, December
SMITH, E. 2007, 'Using a scenario approach', in Hopkins L.D., Zapata M., ed., *Engaging the Future, Forecasts, Scenarios, Plans, and Projects*, Cambridge, MA: Lincoln Institute of Land Policy
SMITH MORRIS, E. 1965, 'New Urban Design Concepts, Greenways and Movement Structures: The Philadelphia Plan', in Lewis D. (ed.) *The Pedestrian in the City*, London: Elek Books
SMITHSON, A. 1968, ed., *Team 10 Primer*, London, Cambridge, MA: MIT Press (reprint of the 1962 edition with new preface)
SOHN, E. 2007, 'Organicist Concepts of City Landscape in German Planning after the Second World War', *Landscape Research*, vol. 2, n. 4
SORIA Y MATA, A. 1882, *La Ciudad Lineal*, Madrid: El Progreso
SORIA Y MATA, A. 1984, *La cité linéaire - Nouvelle architecture de villes*, EBA (anastatic reprint of the French translation by Benoit-Levy G. of the report presented by the 'Compañia Madrileña de urbanización' at the *Premier Congrès International de l'art de construire villes et organisation de la vie municipale*, Gand 1913)
SNODGRASS, A., COYNE, R. 2006, *Interpretation in Architecture: Design as a Way of Thinking*, London and New York: Routledge
STEIN, G. 1935, *What is English Literature?* in *Lectures in America*, New York: Random House, republished by Virago Press, 1988
STENGERS, I. 1987, ed., *D'une science à l'autre. Des concepts nomades*, Paris: Seuil
STRAUVEN, F. 1985, *René Braem. Les aventures dialectiques d'un moderniste flamand*, Bruxelles: Archives d'Architecture Moderne
SUDJIC, D. *The 100 Mile City*, Great Britain: André Deutsch
SUTTON, S.B. 1971, *Civilizing American Cities: A Selection of Frederick Law Olmsted's Writings on City Landscape*, Cambridge, MA: MIT Press (from the paperback edition of 1979)
TAUT, B. 1920, *Die Auflösung der Städte oder Die Erde eine gute Wohnung*, Folkwang Verlag, Hagen
TAUT, B. 1963, *1920–1922 Frülicht, Eine Folge für die Verwirklichung des neuen Baugedankens*, Berlin Frankfurt/M Wien: Ullstein
TIETJEN, A. 2009, *Towards an Urbanism of Entanglement*, PhD thesis, Landscape and Urbanism, Aarhus School of Architecture
TOMASI DI LAMPEDUSA, G. 1958, *Il Gattopardo* Milano: Feltrinelli
TZONIS A., LE FAIVRE, L. 1999, 'Beyond Monuments, Beyond Zip-a-tone, Into Space/Time: Contextualizing Shadrach Woods's Berlin Free University, a Humanist Architecture', TU Delft, online publications
UNGERS, O.M. 1976, 'Planning Criteria', in *Lotus International* n. 11

UNGERS, O.M. 1978, 'Die Stadt in der Stadt. Berlin Das Grüne Stadtarchipel', with P. Riemann, H. Kollhoff, A.A. Ovaska, R. Koolhaas. *Lotus International* n. 19

VAN DEN HEUVEL, D. 2000, 'The Diagrams of Team 10', *Daidalos* n. 74

VAN DER HELM, R. 2005, 'The future according to Frederik Lodewijk Polak: finding the roots of contemporary future studies', in *Futures* n. 37

VELTZ, P. 1996, *Mondialisation, villes et territoires: une économie d'archipel*, Paris: PUF

VENTURI, R., SCOTT BROWN, D., IZENOUR, S. 1972, *Learning from Las Vegas*, Cambridge, MA: MIT Press

VERNE, J. 1879, *Les 500 millions de la Bégum* (from the edition of 1966, Paris: Le Livre de poche)

VIDAL DE LA BLACHE, P. 1903, *Tableau de la géographie de la France* (photographic reproduction, 1979, Paris: Librairie Jules Tallandier)

VIDAL DE LA BLACHE, P. 1922, *Principes de Géographie humaine,* published posthumously starting with manuscripts of the author, by Emmanuel de Martonne, Paris: Librairie Armand Colin

VIDLER, A. 1977, 'The Third Typology', in *Oppositions, Selected Readings from a Journal on Ideas and Criticism in Architecture, 1973–1984*, 1998, New York: Princeton University Press

VIDLER, A. 2000, 'Diagrams of Utopia', *Daidalos* n. 74

VIEILLE BLANCHARD, E. 2007, *Croissance ou stabilité? L'entreprise du Club de Rome et le débat autour des modèles*, in Dahan Dalmedico A., ed., *Les modèles du futur*, Paris: La Découverte

VIGANÒ, P. 1994, *Il progetto come descrizione*, in *Quaderno della ricerca sulle trasformazioni dell'habitat contemporaneo,* n. 2, Venezia: IUAV

VIGANÒ, P. 1995, 'Progetto come descrizione 2', paper presented at the 2nd International Conference of Urbanism *Descrivere il territorio*, Prato, unpublished manuscript

VIGANÒ, P. 1999, *La città elementare*, Milano: Skira

VIGANÒ, P. 2000a, *Il progetto urbanistico: temi di ricerca e futuro della città*, in Piroddi E., Scandurra E., De Bonis L., ed. *I futuri della città. Mutamenti, nuovi soggetti e progetti*, Milano: Franco Angeli

VIGANÒ, P. 2000b, 'Forme progettuali e tipi di rappresentazione', in Bergamo, *Il Progetto del Piano regolatore generale, Urbanistica quaderni* n. 27

VIGANÒ, P. 2001, ed., *Territori di una nuova modernità/Territories of a New Modernity*, Napoli: Electa

VIGANÒ, P. 2003, 'The Design of the Gattopardo', *Hunch,* vol. 6/7

VIGANÒ, P. 2005a, *Un espace de description; Un espace d'hypothèses; Un espace d'abstraction* in AA.VV. *Xland Suisse, Travaux sur le territoire lausannois*, Lausanne: EPFL (works of the design atelier, 2003–2004)

VIGANÒ, P. 2005b, *No Vision?*, in De Michelis M., Pakesch P., ed., *MStad/MCity, European Cityscapes* exhibition catalogue, Kunsthaus, Graz

VIGANÒ, P. 2008a, 'Water and Asphalt, The Project of Isotropy in the Metropolitan Region of Venice', *Cities of Dispersal, Architectural Design* Jan./Feb.

VIGANÒ, P. 2008b, 'Le projet comme producteur de connaissance', in Tsiomis Y. *Matières de ville, enseigner le projet urbain*, Paris: Éditions de la Villette

VIGANÒ, P. 2008c, 'Scenari: il progetto come produttore di conoscenza', in Bozzuto, P., Costa, A., Fabian, L., Pellegrini, P. 2008, *Storie del futuro, Q4*, Roma: Officina

VIGANÒ, P. 2008d, 'Inventori di paesaggi / Inventors of landscapes', in AA.VV., *Ereditare il paesaggio*, Milano: Electa

VON SEGGERN, H., WERNER, J., GROSSE-BÄCHLE, L. 2008 ed., *Creating Knowledge. Innovation Strategies for Designing Urban Landscapes*, Berlin: Jovis Verlag

WAGNER, M. 1929, *Städtebauliche Probleme in amerikanischen Städten und ihre Rückwirkung auf den deutschen Städtebau*, special issue of *Deutsche Bauzeitung*, Berlin

WAKEFIELD, E.G. 1849, *A View of the Art of Colonization,* (reprinted in 2001, Kitchener, Canada: Batoche Books Limited)

WELLS, H.G. 1901, *Anticipations of the Reaction of Mechanical and Scientific Progress upon Human Life and Thought*, London: Chapman & Hall, LD.

WELLS, H.G. 1902, *The Discovery of the Future* (from the edition of A.C. Fifield, London, 1913)

WELTER, V.M. 2002, *Biopolis: Patrick Geddes and the City of Life*, Cambridge, MA: MIT Press

WEST CHURCHMAN, C. 1967, 'Wicked Problems', in *Management Science* vol. 14, n. 4

WHITE, H. 1973, *Metahistory: The Historical Imagination in Nineteenth-Century Europe*, Baltimore: Johns Hopkins University Press

WHITE, H. 1978, *Tropics of Discourse: Essays in Cultural Criticism*, Baltimore: Johns Hopkins University Press

WOODS, S. 1960, 'Stem', *Architectural Design* n. 5

WOODS, S. 1961 'Stem', *Carré Bleu* n. 3

WOODS, S. 1962, 'Web', *Carré Bleu* n. 3

YOURCENAR, M. 1936, *Comment Wang-Fô fut sauvé, La Revue de Paris* n. 44, 1 (15.2.36)

ZAITZEVSKY, C. 1982, *Frederick Law Olmsted and the Boston Park System*, Cambridge, MA and London, England: Belknap Press

ZARDINI, M. 1996 ed., *Paesaggi ibridi. Un viaggio nella città contemporanea*, Milano: Skira

ZARDINI, M. 2005 ed., *Sense of the City An Alternate Approach to Urbanism*, Baden: Lars Müller Verlag

Monographic issues of magazines

ANY n. 23, 1998: *Diagram Work*

Beyond n. 1, 2009: *Scenarios and speculations*

Daidalos n. 74, 2000: *Diagrammania*

El Croquis n. 79, 1996: *OMA/Rem Koolhaas*

El Croquis n. 94, 1999: *Neutelings Riedijk 1992–1999*

Fisuras, July 2002

Les Annales de la recherche urbaine, n. 67, 1995, *Densité et espacements*

Lotus International n. 127, 2006

Oase n. 44, 1998

Encyclopedias, Dictionaries

Enciclopedia Einaudi

Vocabulaire européen des philosophies, 2004, Seuil – Le Robert

Enciclopedia Garzanti di filosofia, 1961, Garzanti Editore

All the sketches, unless otherwise indicated, are by P. Viganò.
The redrawings of the diagrams are by P. Viganò with Studio Secchi-Viganò (L. Domenichini, S. Dussin, A. Pagnacco, C. Pregazzi) and the students of the course in "Urban Planning and Technique" that I have taught in these years at the IUAV University of Venice.
In particular:
- M. Baldassa, redrawing of *No-Stop City*, 1969-1972, Archizoom Associati, pp. 26;
- S. Tropea, deconstruction of the linear city of Soria y Mata, pp. 64 and pp. 66;
- M. Voltini, two principles of linear organization, pp. 67, and examples of linear city projects, pages 68 and 69;
- V. Maffessanti, deconstruction of the diagram of the *Garden City* by E. Howard, pp. 70;
- L. Zandigiacomi, reinterpretation of the "Green City" project for Moscow, 1930, pp. 76.
The images on pp. 220 are part of the research project "Water and Asphalt, The Design of Isotropy in the Metropolitan Area of Venice", Ricerca Prin, 2007-2008 (B. Secchi, P. Viganò, L. Fabian, P. Pellegrini).
The map of the areas below the level +5 m. on pp. 276 is part of the research *The Extreme City* (P. Viganò, L. Fabian e C. Cavalieri); source: Document Actions (European Environment Agency).
The model on pp. 276 is part of the work of A. Bonadio, A. Curtoni, G. Mazzorin, I. Salmaso for the workshop FSE 2009, IUAV University of Venice, "Climate Change: Scenarios for New Territories" (directed by P. Viganò).
The images on pp. 334 and pp. 342 were developed in the Ph.D workshop *Scenarios for the European City and Territory*, IUAV University of Venice, 2002, and taken from Viganò P., ed., 2004, *New Territories*, Roma: Officina Edizioni.
In particular:
- P. Aureli, A. Capurro, A. Costa, M. Giliberti, J. Soares, A. Tegaldo, "Euro_tube", pp. 334; "isotropic Europe", "increased archipelago", "exacerbated differences", pp. 342;
- S. Basso, S. Bencini, S. Escarria, E. Lancerini, G. Mantia, "slow territories", pp. 342;
- P. Bozzuto, L. Fabian, S. Loddo, G. Musante, A. Scarponi, D. Tiozzo, "X-treme Europe", pp. 342.
The photographs of the diffuse city on pp. 174 and pp. 216 are by P. Viganò.
The photograph of the Cemetery of Kortrijk, on pp. 226, is by Syb'l S.- Pictures, Carine Demeter.